THE
AR-15/M16
A PRACTICAL GUIDE

THE
AR-15/M16
A PRACTICAL GUIDE

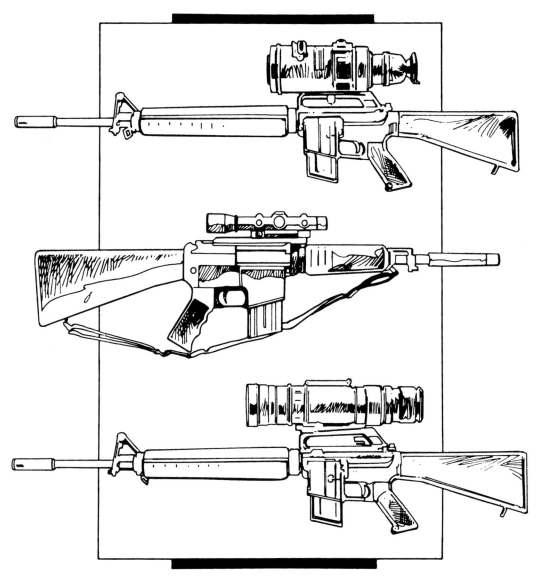

DUNCAN LONG

PALADIN PRESS
BOULDER, COLORADO

Also by Duncan Long:

AK47: The Complete Kalashnikov Family of Assault Rifles
AR-15/M16 Super Systems
Combat Rifles of the 21st Century
The Complete AR-15/M16 Sourcebook
Hand Cannons: The World's Most Powerful Handguns
The Mini-14: The Plinker, Hunter, Assault, and Everything Else Rifle
Mini-14 Super Systems
Modern Combat Ammunition
Modern Sniper Rifles
The Poor Man's Fort Knox: Home Security with Inexpensive Safes
Ruger .22 Automatic Pistol: Standard/Mark I/Mark II Series
Streetsweepers: The Complete Book of Combat Shotguns, Revised and Updated
The Sturm, Ruger 10/22 Rifle and .44 Magnum Carbine
Super Shotguns: How to Make Your Shotgun Into a Do-Everything Weapon
The Terrifying Three: Uzi, Ingram, and Intratec Weapons Families

The AR-15/M16: A Practical Guide
by Duncan Long

ISBN 10: 0-87364-321-6
ISBN 13: 978- 0-87364-321-4
Printed in the United States of America

Published by Paladin Press, a division of
Paladin Enterprises, Inc.
Gunbarrel Tech Center
7077 Winchester Circle
Boulder, Colorado 80301 USA
+1.303.443.7250

Direct inquiries and/or orders to the above address.

Visit our Web site at www.paladin-press.com

Contents

ACKNOWLEDGMENTS

It is impossible for me to extend my thanks to all the individuals and companies that have aided me in writing *The AR-15/M16: A Practical Guide.* I thank them all, with a special thank you to Paul Fred Long for helping with a few of the more important photographs.

And, of course, a very special thanks to Maggie and Kristen.

Warning

1. History and Development

Probably no small arm has had so much written about it—both positive and negative—as the AR-15. Later designated the M16 by the U.S. military, the AR-15 has been praised and hated throughout its existence.

The AR-15 was originally designed to take advantage of modern industrial methods which would make it possible to produce the weapons quickly without large numbers of special milling operations. At the same time, modern industrial machinery allowed tight enough tolerances to allow ready substitution of parts when repair or replacement of pieces was necessary. The rifle was designed to be as light and handy as possible; it was chambered in a caliber which allowed for a very light recoil while still shooting a bullet which took advantage of the high-velocity wounding characteristics of modern military small arms.

As one might imagine, such a weapon has styling and features which make it look far different from a hunting rifle or older military rifles. The "look" and design features of the AR-15 rifle have quickly caught on, however, so that now most modern "assault rifles" developed for the militaries of the world copy many of its features, with many countries making outright copies of the rifle for their own use.

Like the rifle, the 5.56mm (.223 Remington) round developed for the rifle has had a great influence on military thinking. Many countries of the Free World have adopted the round for their battle rifles, while the USSR has switched to a round for their assault rifles which is similar ballistically (though somewhat inferior) to that introduced to the battlefield by the AR-15.

The AR-15 was developed late in the 1950s by the Armalite Division of Fairchild Engine and Airplane Corporation. Rather than using the stock and rifle designing style that was common to the U.S. military's modern weapons (such as the M1 Garand or M-14), the company departed radically toward a style more like that of the assault rifle ("Sturmgewehr") of the German army. That rifle was developed during the end of World War II as the MP43 (which in turn had a series of changes which led to other modern rifles like the CETME and G3/Heckler & Koch 91).

Toward the end of the Second World War, the Germans had been working with several versions of short, lightweight rifles that fired an intermediate round that struck a balance between the pistol and rifle rounds of the time. Weapons seized at the end of the war greatly impressed Allied military leaders, and work was started toward a modern "assault rifle" which would incorporate the ideas developed by the Germans.

Unfortunately, while the USSR was creating a very efficient rifle and round that took advantage of the forward-looking German idea, the U.S. military decision makers refused to give up the powerful, long-range rifle round, even after numerous studies showed that the WW I-WW II vintage round was too powerful for normal combat needs. Because such a powerful round made it impossible to develop a small, lightweight weapon without excessive recoil, designers of military rifles were virtually locked into creating a heavy rifle to go with the round.

While the U.S. military was trying to create a weapon which could use large, powerful rounds, Armalite (a commercial company) was trying to look forward to what would be needed on the world's gun market in the future.

To see how the AR-15 developed, it is necessary to look at some of the other rifles which Armalite was engaged in designing.

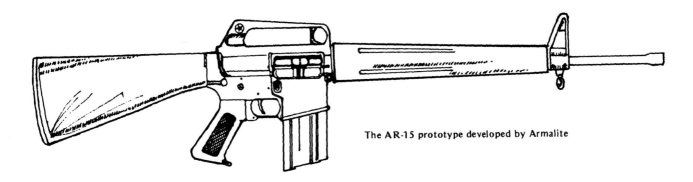

The AR-15 prototype developed by Armalite

OTHER AR RIFLES

Work had started on the AR-1 (Armalite Rifle number 1) in 1947 and completed in 1954. The result was Armalite's "Parasniper Rifle" which was a scoped, bolt-action sporting rifle which could double as a military sniper rifle.

The AR-1 was chambered in .308 Winchester (7.62mm NATO) and incorporated two features that would be seen in later rifles of the series: a fiberglass stock (filled with foam) and an aluminum barrel with a steel lining.

Though the rifle never got beyond the prototype stage, it did show the possibilities that could be realized in creating a rifle that took advantage of modern materials and techniques.

On October 1, 1954, Armalite became a part of the Fairchild Engine and Airplane operation. Armalite was now given the go-ahead to develop firearms which would incorporate the advantages in speed and economy that could be realized in producing rifles using modern materials (plastics and alloys) and newer manufacturing techniques rather than the older, labor-intensive methods then employed in building most conventional rifles.

Eugene Stoner joined Armalite in 1954 and brought with him a semiautomatic rifle he had been working on. This rifle became the AR-3.

The AR-3 had an aluminum body and a fiberglass stock. Though it, too, never got beyond the prototype stage, it did demonstrate the practicality of the goals of Armalite, and many of its features were to be found on subsequent rifles.

Several other talented workers were soon teamed up with Eugene Stoner at Armalite at this time: L. James Sullivan (who worked as a designer/draftsman) and Robert Fremont (who supervised prototype manufacture and led studies which determined whether the tolerances needed for rifles would be practical from a mass-production standpoint). These three men worked on a number of the Armalite weapons.

It should be noted that a number of the Armalite firearms were being developed during the same period of time. At any one time it was possible that several weapons were in various stages of development. Thus the Armalite designation numbers of "AR-1," "AR-10," etc., do not necessarily reflect the chronological relationship in terms of development, nor do they show the order in which the firearms were introduced to the public.

Although the original goal of Armalite was to produce sporting arms, their first success came in 1957 with the AR-5 rifle, which was designed to conform to the U.S. Air Force's requirements for a survival rifle to be carried by air crews.

The AR-5 was a bolt-operated rifle chambered for the .22 Hornet. The rifle used a detachable magazine and a barrel which was held to the front of the receiver by a threaded ring; it was 30-1/2 inches long when assembled and 14 inches when broken down. The rifle receiver/action, barrel, and magazine could all be stowed in the hollow fiberglass stock when the firearm was broken down for storage. The materials used to make the rifle were so lightweight that the rifle would actually float because of the buoyancy of the hollow stock (a strong selling point for a survival rifle which might conceivably see use in a life raft or near the water).

The AR-5 was designated the MA1 by the U.S. Air Force, but Armalite never saw any great monetary results from the rifle as the Air Force had a large inventory of M4 and M6 Survival Guns which prevented the purchase of any large numbers of the AR-5 (MA1).

To take advantage of the work they'd done on the AR-5, Armalite created the AR-7 for the commercial market. The rifle was chambered for the

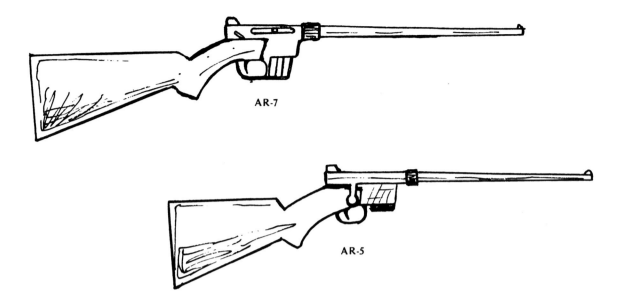

AR-7

AR-5

popular .22 LR. The new rifle was changed to a semiauto blow-back action (which was inexpensive to manufacture), and the aluminum barrel (with steel lining) was lengthened to 16 inches to conform to BATF regulations for regular rifles. The ability of the rifle to be broken down and stored in the hollow stock was retained, as well as the rifle's ability to float on water.

The AR-7 was quite successful commercially. Rather than continue production on the rifle, Armalite sold the rights to it to Charter Arms Corporation.

The AR-9 was a semiautomatic aluminum (barrel and body) shotgun which incorporated a number of design features which later found their way into the AR-10 and AR-15 rifles (including its rotating bolt design). Rather than market the 5-1/2 pound shotgun, Armalite decided to shelve it in 1955, choosing to use many of its features in developing the AR-17 shotgun.

Development of the AR-10 was started in 1953 before Stoner joined Armalite. The AR-10 was originally chambered for the .30-'06 cartridge. It was modified two years later for the new 7.62mm NATO cartridge, which appeared to be on its way to becoming the standard round for much of the Free World.

The AR-10 was the first of the Armalite rifles to use a gas system which did not have a gas piston to unlock the firearm's chamber when a round was fired.

A version of the AR-10 was submitted to the U.S. Springfield Armory in 1956 for testing as a possible replacement for the M1 Garand rifle. Unfortunately, the rifle had a titanium barrel surrounded by an aluminum jacket (similar to that developed for other of the Armalite weapons). Although AR-10 was able—unlike the M-14—to shoot in the automatic mode while remaining easy to control (due to its straight back design and a special titanium muzzle brake), the barrel burst during testing, and the rifle was therefore disqualified.

Additionally, the muzzle brake was expensive (since titanium is rather rare and has to be imported to the United States), and the rifle's

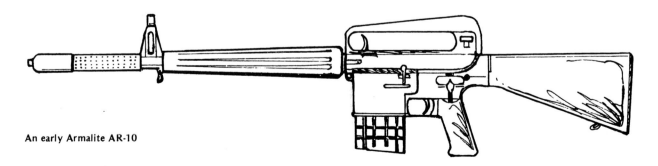

An early Armalite AR-10

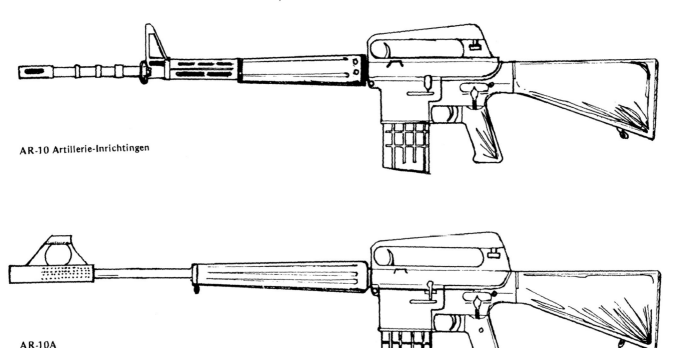

AR-10 Artillerie-Inrichtingen

AR-10A

"Buck Rogers" look undoubtedly had some negative effects with conservative forces in the military.

Stoner—with the assistance of the Springfield Armory—developed an all-steel conventional-style barrel for the rifle, but the damage had already been done. The U.S. Army chose the M14 rifle over the FN FAL and AR-10. Time would prove that they had chosen the inferior weapon.

Though the rifle was still being redesigned by Stoner and others, Fairchild had actively promoted the AR-10 and licensed the government-owned arsenal of Artillerie-Inrichtingen of Hembrug, Holland, to manufacture the rifle.

This was unfortunate because the state-controlled arsenal was not able to tool up in time to meet the demand for the AR-10 which Fairchild's aggressive publicity campaign had created.

Because of the problems created by the tardy Holland arsenal, a number of inferior—but available—types of rifles were chosen by many countries to fill the need for a military rifle chambered for the 7.62mm NATO. Thus, only 5,000 AR-10 firearms were manufactured by the Artillerie-Inrichtingen. By the time the arsenal was ready to manufacture large quantities of the rifles, the interest in the firearm had subsided.

Production of the rifle by the Artillerie-Inrichtingen was finally halted in 1959, and Colt's Patent Firearms was licensed to manufacture the improved version of the AR-10, designated the AR-

10A. (Major improvements of the AR-10A were a stronger extractor, a more reliable magazine system, and a cocking handle toward the rear of the receiver.)

In addition to a short-barreled carbine version of the AR-10, a number of modified AR-10s were also created in an effort to capture some of the market which seemed to want a new LMG (Light Machine Gun) using the 7.62mm NATO round. To these ends, a bipod was added to the AR-10, and it was modified to use belt-feed ammunition. Later, it was mounted on a tripod and the gas tube was modified and spring loaded for use with quick-change barrels. None of the variations attracted much interest among military buyers, however, so Colt decided not to produce the AR-10A. Rather, the company started marketing the AR-15 after securing the rights to it.

The AR-10 almost got another lease on life as the U.S. Army's sniper rifle. A modified AR-10 was one of six rifles tested at the Aberdeen Proving Ground in 1977. The rifle was modified by the Rock Island Arsenal; major modifications consisted of the removal of the front sight assembly and the rear sight/handle and the placement of a scope base and ART scope on the rifle. But the tests were inconclusive, indicating only the need for a better ranging system and more accurate ammunition.

Most authorities feel that the AR-10 is an excel-

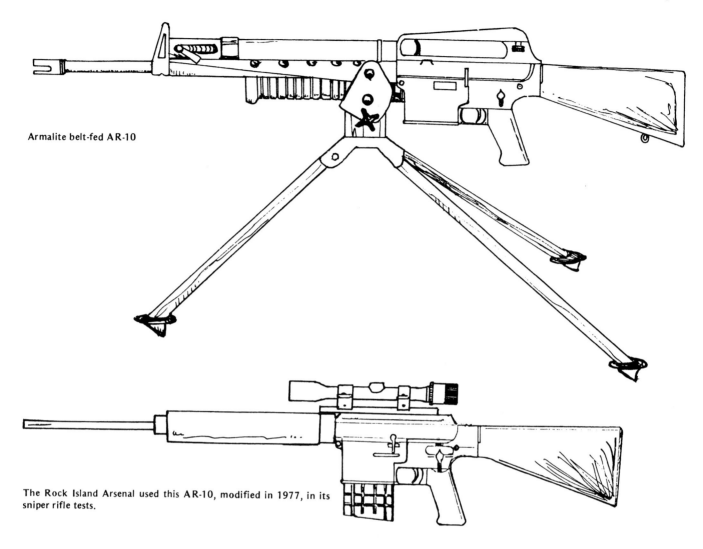

Armalite belt-fed AR-10

The Rock Island Arsenal used this AR-10, modified in 1977, in its sniper rifle tests.

lent weapon which missed its place in history because of poor timing and marketing. (Very possibly the last of the AR-10 has not been seen, however. While work on this book was being done, several small commercial manufacturers in the United States told of plans to manufacture a version of the AR-10 for the public. This might meet with considerable success considering the collectors' demand for the AR-10 and the desires of many shooters for a lightweight assault-style rifle chambered for .308 Winchester/7.62mm NATO.)

The AR-11 was a conventionally stocked rifle which resembled the AR-3. It is of interest in that it used the .222 Remington, which was modified to become the round used in the AR-15.

The AR-12 was a version of the AR-10, modified to make it easy to mass-produce with the use of steel stampings for the upper and lower receiver rather than the cast/machined aluminum receivers. The AR-12 was chambered for the 7.62mm NATO and might have been made at about half the cost

of the AR-10. The AR-12 never went beyond the prototype stage, but it did set the stage for the development of later rifles (such as the AR-16 and AR-18/180).

The AR-14 was the sporting version of the AR-10 with a conventional Monte Carlo stock (no pistol grip) and iron sights. It was chambered for .308 Winchester (7.62mm NATO), .243, and .358.

The AR-16 rifle was produced during 1959 and 1960. It is notable because it made use of the inexpensive manufacturing techniques pioneered by the AR-12. The AR-16 was chambered for the 7.62 NATO/.308 Winchester. Though the AR-16 wasn't successful, it did lead to the AR-18 which would become a competitor with the AR-15 for use among the militaries of the Free World.

THE AR-15 RECOIL/BUFFER SYSTEM

Developed from 1956 to 1959, the AR-15 was designed to make use of a number of the principles and designs found in other weapons, as well as those of other Armalite weapons. Though one might argue that the AR-15—and the AR-10 before it—incorporated little that hadn't been done before, the genius of Stoner in arranging the layout and picking out the best other firearms' features cannot be understated.

Like the AR-10, the AR-15 used a recoil/buffer system in the stock and used the gas from the discharge of a fired round to unlock the chamber and operate the action of the rifle. Rather than use a heavy spring and rod to move the action, the AR-15 has only a light, hollow gas tube over the barrel.

The buffer and gas system made it possible to keep the AR-15 lightweight and gave it a beautiful balance not found in many other rifles. Because the weight of the rifle is centered just in front of the pistol grip, the rifle is much easier to carry than many others, seeming lighter than it actually is.

The gas system was tied into the basic design of the rifle. Stoner realized that for the gas system to work properly, a fast powder would have to be used. If a slow powder were to be utilized, the powder would not be completely burnt by the time it entered the gas system to propel the bolt open and to operate the reloading mechanism. Repeated firings under such conditions would create a major fouling problem in the bolt of the rifle until jamming problems would finally develop. Because of this, one of the design requirements was that the rifle would use fast IMR (Improved Military Rifle) powders.

The AR-15 was originally chambered for the .222 Remington used in the AR-11. The .222 case was used with a special 55-grain boat-tail bullet developed for Armalite by the Sierra Bullet Company. The cartridge case was not large enough to create the velocity desired, so it was lengthened and the round was called the .222 Special. The first

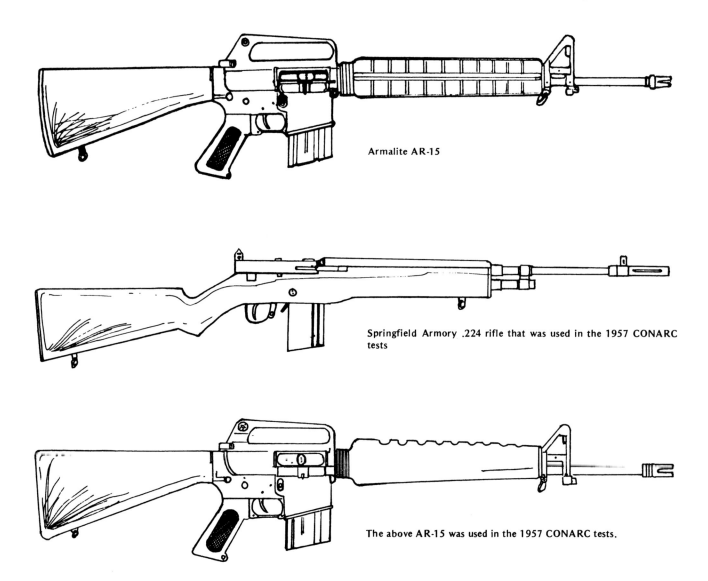

Armalite AR-15

Springfield Armory .224 rifle that was used in the 1957 CONARC tests

The above AR-15 was used in the 1957 CONARC tests.

ammunition manufactured for use in testing was made by Remington and bore the headstamp of ".222 Special." Later, Remington marketed a different round—the .222 Magnum—so Armalite renamed the .222 Special the ".223 Remington" to avoid confusion since the .223 round would fit and fire in rifles chambered for the .222 Magnum—with dangerous results!

TESTING AND ADAPTATION BY THE U.S. MILITARY

The AR-15 was originally developed to meet the potential military market that seemed to be developing in the United States.

The U.S. Continental Army Command (U.S. CONARC) had felt that their current small arms projects were showing a need for a lightweight, small caliber weapon. In 1957, the Infantry Board in Fort Benning had asked Winchester-Western (a Division of the Olin Mathieson Corporation) and Armalite to develop weapons and ammunition to be tested as a possible standard to replace the M-14 and 7.62mm NATO round—both of which had been developed by the military itself—which had been doing poorly. A third rifle was created to compete against the two commercial rifles by the U.S. Springfield Armory.

The requirements set forth by the Infantry Board were as follows:

- The rifle was to weigh less than 6 pounds when fully loaded.
- The accuracy and trajectory would be equal to or better than the M1 rifle out to 500 yards (this was almost twice as long a range as studies indicated would be needed—perhaps a compromise among military leaders).
- The rifle would need to be capable of automatic fire.
- The round the rifle used would be able to penetrate body armor, steel helmet, or a 10-gauge steel test plate out to 500 yards.
- The round would be equal in lethality to the M1 Carbine within 500 yards.
- The weapon would have a detachable, 20-shot magazine.

With these requirements in mind, the AR-15 was created, using many of the design features of the AR-10.

It is interesting to note the atmosphere of the times among military planners. Their thinking had been greatly influenced by various research projects, conducted by the military, which led to the decision to start looking for a new rifle for combat use. The research also dictated the requirements made by the Infantry Board. There were three programs that profoundly changed the thinking of the military concerning the type of small arms needed and how they should be used: ALCLAD, SALVO, and Special Purpose Individual Weapon (SPIW).

Though the ALCLAD study dealt principally with the requirements for better body armor (some of which had proved to be highly effective when tried out in the Korean War), it got into areas of interest to small arms experts when it examined such things as at what range casualties occurred, what happened when the human body was struck by bullets or shell fragments, the frequency and distribution of wounds, etc. To obtain the answers to these questions, a study was made of over three million casualty reports from the first two world wars and the Korean War.

Though it flew in the face of current military thinking, the study concluded that:

- Nearly random shots produced more casualties than aimed fire in combat.
- Rifle fire was seldom used effectively at distances greater than 300 yards.
- Most rifle casualties were produced at ranges of 100 yards or less.
- Even expert marksmen could seldom hit targets beyond 300 yards because of terrain features or the need of cover by the marksmen.

These facts led to the SALVO project in the early Fifties. A number of their findings would also affect later small arms' design. Two of their major points were that lightweight bullets were adequate at normal combat ranges, and long bursts of fire tended only to waste ammunition while three-round bursts were the most effective automatic fire in small arms. The principle thrust of the SALVO project was to outline the requirement for a small caliber rifle capable of automatic fire.

The SPIW project was carried out during the early Fifties. It was principally aimed at producing a weapon which could create a multiple projectile pattern of shots through the use of flechettes packaged in one round, similar to that of a shotgun. The project only led to limited success due to the limited range and large dispersion of the projectiles created by the researchers. The rounds precluded accurate single shots to be fired if there were a need for them, and the weapon was not able to fire out to the 300-yard maximum which was sometimes needed in combat (according to the research). Though the idea of a flechette-loaded round did not meet with acceptance at the time, it did lead to the production of a flechette-loaded "grenade" round for 40mm grenade launchers (which later found their way to versions that would be placed under the barrel of some of the military's AR-15s).

Though efforts were even made to mount single flechettes into saboted rifle rounds, the SPIW program was generally seen as having met with failure. The final outcome was to make military planners stick with conventional rifle-style ammunition rather than the shotgun shell-like rounds created by the SPIW project.

So, after these projects had shown what the military needs were, CONARC decided to ask Winchester and Armalite to submit rifles for testing. On March 31, 1958, ten AR-15s were delivered to the Infantry Board for trial.

During the tests at Aberdeen Proving Ground and Fort Greely, a number of modifications were made to the AR-15:

- The barrel was strengthened (to allow for firing with small amounts of water in the bore of the rifle).
- The cocking handle was moved from the inside of the carrying handle to the back of it.
- The trigger guard was changed to allow it to swing down for cold-weather shooting with mittens.

These changes made the rifle about a pound heavier than the original specifications of the Infantry Board, but resulted in a better rifle as far as the military testers were concerned.

During 1959, the U.S. Army conducted tests involving the AR-15 and Winchester's .224 Lightweight Military Rifle, but reached no decision as to whether or not to adopt the small rounds or the rifles for actual military use. During the tests, the troops involved were very impressed with how the smaller rifles (especially the AR-15) handled and

preferred the smaller caliber weapons over the heavier M-14 rifles. During the tests it became apparent that the M-14 rifles were almost impossible for the average combat soldier to control under actual combat conditions (as opposed to target-style shooting).

The Army's test report concluded that the U.S. Army should develop a lightweight, reliable rifle "like the AR-15" to replace the M-14 and also suggested that the increased firepower afforded by such a weapon would allow the reduction in the then-current squad's size.

Following the tests, Winchester discontinued work on their military rifle, and the management of Armalite sought to divest themselves of the AR-15 (apparently, neither company felt that the U.S. military would really be interested in actually using either of the rifles in the near future).

Colt Firearms Corporation saw the potential for the AR-15 and bought the license for manufacturing it—along with the rights for the AR-10—from Armalite in 1959. Colt's aggressive sales techniques enabled them to sell a number of the rifles to several small Southeast Asian countries. (The rifle was much easier for the average Asian soldier to control since the rifle was lighter and offered less recoil—both important considerations for the smaller physique of the Asian troops.)

During its use in the Asian arena of combat, the lightweight rifle showed how lethal it and the 5.56mm bullet it fired actually were.

Meanwhile the U.S. Army—just as its tests in 1959 had suggested—was finding that the M14 was too heavy for troops to handle easily. The heavy round the M14 fired made it uncontrollable during automatic fire, even with the special E2 stock that was designed for it as somewhat of an afterthought.

In 1960 the U.S. Air Force started its own tests of the AR-15 at Lackland Air Force Base in Texas, hoping to use it as a replacement for the .30 Car-

This Winchester .224 lightweight military rifle was used in the 1957 CONARC tests.

bine. In 1961, the Air Force procured 8,500 AR-15s. That same year, the Army—perhaps recognizing the problems the M14 would suffer in combat use—also purchased 8,500 to test.

In 1962 Colt persuaded the Advanced Research Project Agency (ARPA) of the Department of Defense to test one thousand AR-15s in its Project Agile (which was aimed toward finding a better weapon for use in the Vietnam War).

ARPA found the AR-15 to be very suitable for combat, and this finding led to further studies by the Department of Defense as well as the Army.

ARPA came to the following conclusions:

- A squad armed with AR-15s had five times the level of overall kill potential compared to a squad armed with M-14s.
- The AR-15 could also be produced more cheaply and with a higher degree of quality than the M-14.
- The AR-15 was more reliable, durable, rugged, and easier to care for than the M-14 under the adverse conditions often found in combat.
- Soldiers learned to shoot better and quicker with the AR-15 (as compared with the M-14).
- Three times as many rounds could be carried by a soldier with an AR-15 (as compared with the M-14) when the weight of both the weapon and the ammunition were taken into account.

From 1962 to 1964 a version of the Armalite AR-16 rifle was created which was chambered for the .223 Remington (5.56mm). Designated the AR-18, it is notable because during the mid-Sixties it competed with the AR-15 for the military market.

The AR-18 was marketed after Stoner left the Armalite Company but—according to Burton T. Miller, who was the Vice President of Armalite—Eugene Stoner was one of the people who was responsible for the development of the AR-18. (Many authorities don't realize that Stoner worked on the AR-18.)

Very possibly the AR-18 is a better rifle than the AR-15; but problems—in a manner reminiscent of those Armalite had had with the AR-10—prevented Armalite from securing a market for the AR-18.

According to Burton T. Miller, some tests were conducted by the Army with the AR-18. These—according to Miller—were all but rigged against the AR-18, with test personnel forcing the rifle to be

used with the wrong type of ammunition and defective magazines so that the ammunition wouldn't feed properly. The company later modified the AR-18 for semiauto fire only and designated the semiauto model as the "AR-180." The AR-180 was aimed at the law enforcement and sporting markets in the United States.

Armalite had sold the rights to the Howa Machinery Company of Nagoya, Japan. The Japanese government had started working to end the war in Vietnam and in an effort to bring pressure on those involved with the war—would not grant an export license to Howa for the shipment of AR-18s to any country even remotely involved with the war.

When the U.S. Army tried to secure some AR-18s for testing, the Japanese government refused to allow the rifles out of the country, and the Army therefore continued to use the more expensive—but readily available—AR-15 manufactured by Colt. By the time the Vietnam War was over, the U.S. military was committed to the AR-15 which had pretty well had all the bugs worked out of its design by that time.

In 1962 the U.S. Air Force conducted additional tests with the AR-15 and, after making some minor modifications of their own to the design, General Curtis LeMay felt the AR-15 would make a good weapon for use by the Air Force. One thousand additional AR-15s were obtained from Colt for further testing. The Air Force soon chose the AR-15 as their standard-issue rifle (designating it the M-16), and the Army (apparently after being pressured by President Kennedy and his "whiz kids") purchased limited numbers of the AR-15 rifles for special troops.

One thousand AR-15s were also sent to the Army of the Republic of Vietnam to "try out" in 1962. These rifles had a fantastic record of reliability during their use in Vietnam: the 1,000 rifles sent to the Army of the Republic of Vietnam—after firing an estimated 80,000 rounds—needed to have only two parts replaced! (Later testing by the U.S. Air Force proved the reliability of the AR-15—with IMR powder—to be very high. During one test in which twenty-seven of the rifles were fired with 6,000 rounds apiece, the malfunction rate averaged only once per 3,000 rounds fired while the part breakage was only once per 6,200 rounds fired!)

There were some political problems, however, that had to be overcome before the U.S. Army was

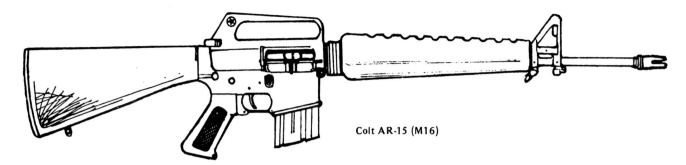

Colt AR-15 (M16)

to accept the AR-15 rifle. The major problem was that the M14 rifle had been produced by the Army for the Army, and a number of people had vested interests in keeping the M14 rifle as the standard weapon.

The M14 was originally developed by Army personnel as a replacement for the Garand, the .30 Carbine, the M3 "Grease Gun," and the Browning Automatic Rifle. As such, specifications for the rifle called for it to fire both in a semiautomatic mode *and* in full automatic. The original "Army Ground Forces Equipment Review Board Preliminary Board Study" called for a 7-pound weapon and a .30 caliber round. This weight/caliber combination might have been practical with a low-powered round like that of the M1 Carbine, but it proved to be impossible with the long-range round that the military felt it needed.

In developing the rifle, the U.S. Army virtually ignored all the United States and English military studies which showed that the battle rifles from World War I through the Korean War had too much power for the job they were being called upon to do.

Flying in the face of reason, Army planners created the T65 cartridge. The T65 round was a shortened version of the old .30-'06 round. The round was not bad for a rifle but was too powerful for a lightweight automatic weapon—one of the prime requirements set up by the Army board. (To make matters worse, the United States later browbeat other NATO countries into accepting the T65 round as the standard round for NATO—the 7.62mm NATO.)

The T44 rifle was developed for the round. The rifle was actually a modified M1 Garand rifle that had been developed shortly after World War II.

Needless to say, there were some problems: The rifle was too light for good control during automatic fire, and the system itself was hard to manufacture because of extremely close tolerances in the gas system and the rifle.

Problems mounted as tests of the rifle and—to make the rifle look good—tests against other contenders were all but rigged. When the rifle was finally fielded, the explosion of several rifles during troop training exercises pretty well finished off any good points the rifle had as far as a lot of those involved with the firearm were concerned.

The new rifle—designated the M14—was supposed to be a lightweight replacement for the Browning Automatic Rifle as well as a standard rifle. The M14 was neither lightweight (12.75 pounds) nor easily controllable in the automatic mode thanks to the recoil of the 7.62mm round. Large numbers of the rifles were actually issued without the automatic fire selector in order to keep the troops from using the uncontrollable mode.

Military designers tried to create a new rifle stock—with a pistol grip and straight back design—and a muzzle brake for the M14 Rifle. The new rifle was first designated the M14E2 and later accepted as the M14A1. (Critics were quick to point out that the "E2" stock made the M14 look remarkably like the AR-10, FN FAL, and AR-15 that had been rejected by the Army so it could continue to use the M14!)

In theory, these modifications made the rifle more controllable in automatic fire. Still, most shooters still couldn't control the 12.75 pound rifle in the automatic mode. The 7.62 was just too much for the M14.

Internal changes were also being made in the M14 from the time it was first being produced so that it gradually became easier to manufacture, and greater design tolerances made the gas piston less apt to lock up and bend if it became fouled.

The changes, however, had been made too late, and many individuals were convinced by now that the Army had chosen too large a caliber for their rifle. As an added mark against the M14, it had turned out to be harder and more costly to mass-produce than had been expected, while the AR-

15—even in limited quantities—was much cheaper than the M14 to manufacture.

But—for those who thought the M14 was a good rifle—all was not lost.

The Army Materiel Command conducted tests comparing the AR-15 to the M-14 during this period and found that the M-14 was *superior* to the AR-15!

Since glowing reports were coming back to the United States about the use of the AR-15 in Vietnam, the Secretary of the Army had the Army's Inspector General look into the tests.

The Inspector General found that the Army Materiel Command (which had a vested interest in the M-14 since it had helped develop that weapon) had rigged the tests by hand-picking target grade M-14 rifles, while the AR-15s were chosen at random without consideration as to how well they shot. The Inspector General also reported that the group had conducted a "dry run" to see how the two models of rifle did, then conducted an "official test" which would not include the parts of the dry-run test in which the AR-15 outperformed the M-14.

USE OF THE AR-15 IN VIETNAM

Despite the results of the rigged test, many of the AR-15 rifles had been purchased by the military. By 1963, many started finding their way into the hands of troops in Vietnam just as the fighting there escalated.

The U.S. Army purchased 85,000 AR-15 rifles for the Green Berets (Special Forces), and Army Airborne units, with the CIA purchasing a large number as well. Nineteen thousand additional rifles were also procured for the U.S. Air Force. (During this period the rifle was called the M16 since it had been thus designated by the Air Force.)

Robert McNamara, then the Secretary of Defense, was impressed with how well the AR-15 worked on the battlefield. Late in 1963, he made the Army the official procurer of the weapons to simplify the paperwork involved in obtaining large numbers of the rifle.

The Army purchased 35,000 additional rifles in 1964, and General W. C. Westmoreland, the U.S. Army Commander in Vietnam, was so impressed with the effectiveness of the weapon that he made an urgent request for additional weapons for the infantry troops in Vietnam. The request was approved, and 100,000 more AR-15s were purchased in 1965 and another 100,000 the following year for use in Vietnam.

While all American Air Force units in Vietnam were armed with the AR-15, most of the rifles purchased by American ground forces went through a trickle-down process which started with the Army's Special Forces and Airborne troops. As more rifles became available to all branches of the military, the AR-15 slowly filtered down to Marine and helicopter crews, the 1st Cavalry Division, and the 173rd Airborne Brigade. As more and more AR-15 rifles were purchased, they finally made their way into the hands of the "grunts" in Vietnam.

A number of problems were blamed on the AR-15 rifle because of poorly designed magazines during the Sixties. The original magazines, made of steel and "waffle patterned" (similar to that of the AR-10), functioned very poorly. These 20-round magazines were soon replaced by aluminum 20-round magazines that weren't much better. The first 20-round aluminum magazines that reached U.S. troops could be jammed if an extra round was pushed into the magazine. (Hence, the rule of only putting eighteen or nineteen rounds into them. That way, if a GI lost count, he would probably only get one round in and still wouldn't jam the magazine.) Needless to say, the few magazines that did jam did little to help the image of the AR-15.

This period of time also saw the fielding of plastic 20-round magazines that were designed to be shipped prepacked with ammunition, used once or twice, and then discarded. These were issued to the Special Forces and Ranger units for evaluation in the field during combat in Vietnam. The disposable magazines were later discontinued as improved aluminum magazines became available. (The idea of plastic magazines has since been adopted by other countries for their assault rifles; one style was created for the AR-15 by Israel which used the AR-15 as their standard rifle until it was replaced by the Galil.)

ARMY CHANGES IN THE AR-15 DESIGN

At this point, the U.S. Army broke the cardinal rule of weapons design: if it works, don't fix it. They proceeded to create a number of minor changes in the AR-15 rifle.

Possibly the worst change was the addition of the bolt forward assist. The AR-15 had been designed to prevent pushing the bolt forward on a round which chambered poorly. That way, if an

The old-style (round) bolt assist

The tear-drop style bolt assist

oversized round caught at the mouth of the rifle's chamber, a soldier would be forced to extract the round and chamber another one. Older rifles had the capability of having the bolt forced forward with the cocking lever, often jamming the round halfway into the chamber. This could make it impossible to fire the weapon or extract the oversized round to chamber a good one.

Nevertheless, the bolt assist was demanded by the U.S. Army. (The original forward assist knob was round. This was later changed to a "tear-drop" shape which allows more clearance between the charging handle and the forward assist.)

Despite the protests of the Marines and the Air Force (who stated that none of the problems encountered during testing could have been overcome by the forward assist modification), the bolt assist was added to the rifles slated for the Army. The Air Force continued to purchase the AR-15 rifles without the "improvements" which the Army demanded.

The handguards were also modified during this time from the ribbed variety (which were retained on the shortened "submachine gun" style AR-15) to a weaker and less comfortable "beaver tail" style, which had a tendency to crack when dropped on a hard surface.

Research conducted by the U.S. Ballistics Laboratory showed that the bullets fired from the AR-15 in Arctic conditions were not stable, and long-range shooting consequently suffered. Despite the fact that the United States was fighting in a warm climate at the time, there was a possibility that the American forces might be called upon to fight in colder climates at a moment's notice. This led to the third modification of the AR-15: a changing of the twist of the barrel from the one twist to fourteen inches of barrel to one in twelve. The faster twist created a more stable bullet and decreased its ability to wound. Coupled with a change to a shorter bullet, many feel the weapon's lethality was decreased by the faster twist.

The fourth change was perhaps the most damning to troops in the field. In June 1963 the Army decided that the IMR powder for which the AR-15 was designed wouldn't maintain the bullet's maximum velocity if fired in Arctic temperatures. Because there was some problem in manufacturing the quantities of ammunition needed, the Army demanded that slower-burning ball powder be used in loading the 5.56mm round. The ball powder had been used in other rifles in the past without prob-

lems and—on paper at least—created a rifle that could be used equally well in Arctic areas or in the heat of a jungle.

For the small advantage gained in using the ball powder, two very disastrous disadvantages were to be encountered. First of all, the ball powder was slow-burning, which meant that the burning particles could enter the gas tube of an AR-15 and move down into the chamber/bolt area of the rifle. Gradually, deposits would result which would cause chambering problems. When coupled with the forward assist and the poor fire discipline among the troops, the chambering problems could jam a rifle so that it became a piece of junk in the field.

The second problem was that the recoil created by the ball powder increased the cyclic rate of the rifle to the point where it could not extract and chamber rounds consistently.

Because of the central procurement policy of the U.S. military, the Army's decision to use the ball powder forced the Air Force to follow suit, since they used whatever ammunition the Army specified for the AR-15 in the Army arsenals.

Thus, by the time the AR-15 had been "improved" by the Army, it had gone from being very reliable in the field to a weapon plagued with constant jamming and breakage.

The original AR-15 used by the Air Force had the designation of "M16," while the modified Army AR-15 was designated the "XM16E1" (the "XM" showing that the rifle was still considered by the Army as "eXperiMental," while the M14 was the standard Army rifle; the "E1" meant that the rifle had undergone some major changes since the introduction of the Air Force's M16).

TESTING

In 1965 Colt conducted its own tests of the AR-15. In those tests, the AR-15 was not likely to fail using IMR powder. When ball powder was used, however, the number of malfunctions would probably be on the order of 50 percent for any given group of AR-15s (i.e., if fired for a period of time, half the rifles would malfunction at least once during any test).

During other tests in 1965 the U.S. Frankford Arsenal found that the AR-15 had 3.85 malfunctions or stoppages per 1,000 rounds fired with cartridges using IMR powder, while the same rifles produced 23.7 failures per 1,000 rounds fired when using ball powder.

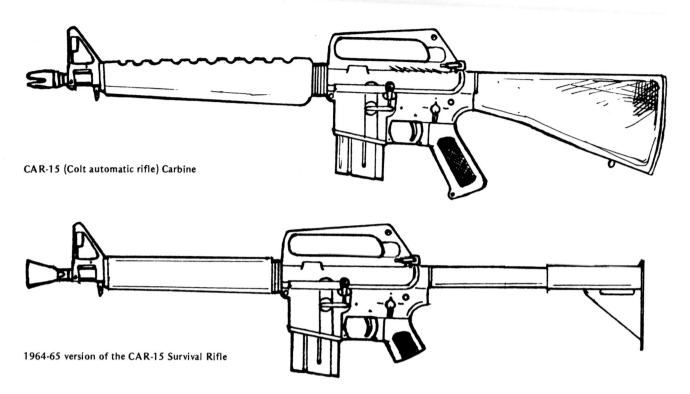

CAR-15 (Colt automatic rifle) Carbine

1964-65 version of the CAR-15 Survival Rifle

THE CAR-15

After buying the AR-15 from the Armalite Corporation, Colt renamed the AR-15 the CAR-15 (Colt Automatic Rifle-15), and experimented with a number of new configurations of the rifle to appeal to a wider military market.

One of the results of the experimentation with new rifle configurations came in 1965 with the first shortened AR-15, produced with a telescoping stock and a short, 10-inch barrel. This rifle weighed 5.3 pounds empty and was designed to be a replacement for the submachine gun. (It is sometimes described as a submachine gun even though it uses rifle ammunition rather than the pistol ammunition used in a true submachine gun.) This short AR-15 made its way to the Special Forces in Vietnam for use and evaluation. Unlike the later versions of the Colt "submachine gun," this one had a handguard and stock which—when the stock was in its short position—made it look like a regular rifle, somehow scaled down for a small child. This version of the AR-15 rifle was 26 inches long with the stock closed and 28.7 inches long open, and cycled at 800 to 850 rounds per minute. The release that allowed the stock to be telescoped was in the buttstock.

This weapon had a very handy length, but the short barrel produced several problems. Since the regular-length AR-15 was designed to have its pow-

der burned up when it reached the gas tube, the shortened barrel of the submachine gun CAR-15 created excessive blast when unburned powder exited the barrel. This reduced the velocity of the bullet to 2,750 feet per second rather than the normal 3,000-plus feet per second of the regular rifle (which wasn't too great a problem at close ranges where a submachine gun is normally employed). The worst problem, however, was that the unburned gases entered the gas tube and caused excess fouling in the chamber area. This—coupled with the faster cycling of the action because of the short gas tube—meant that if the gun weren't cleaned religiously, the rifle would often suffer jamming problems.

A special flash/noise suppressor was designed for the short-barreled CAR-15 which greatly improved the operation of the weapon. (The suppressor was efficient enough that the Federal Bureau of Alcohol, Tobacco and Firearms [BATF] decided the suppressor should be designated as a firearms silencer. Because of this, most of the original "submachine gun" suppressors were destroyed by the U.S. government after the close of the Vietnam War.)

In 1965 Colt also introduced a 29-inch survival rifle which was designed to be taken down into two parts and stored in two parts for emergency use on Air Force planes. Although the stock on the 4.75 pound rifle appears to be telescoping, it in fact does not since space was saved by taking the rifle apart and by its short barrel. A standard issue pistol grip was cut to a shorter length to make it fit into the tight space provided for it. The gun was to have been stored with one 20-round magazine in place in the rifle with three more 20-round magazines in reserve.

THE COLT "COMMANDO"

Using the lessons learned with these two cut-down versions of the CAR-15, Colt came out with the Colt Commando "submachine gun" in 1966.

The Air Force used this rifle without the forward assist on the receiver (just as their AR-15 rifle didn't have a forward assist). The Air Force version was designated the XM177 while the Army's version with forward assist was designated the XM177E1 (the E1 meaning that it had undergone some major changes since the introduction of the earlier versions of the "submachine gun" version; the "XM" stood for "eXperiMental;" and the "177" was used to take it out of the M16, XM16E1, M16A1, etc., line of development numbers to prevent confusion).

The author is shown here with a Commando.

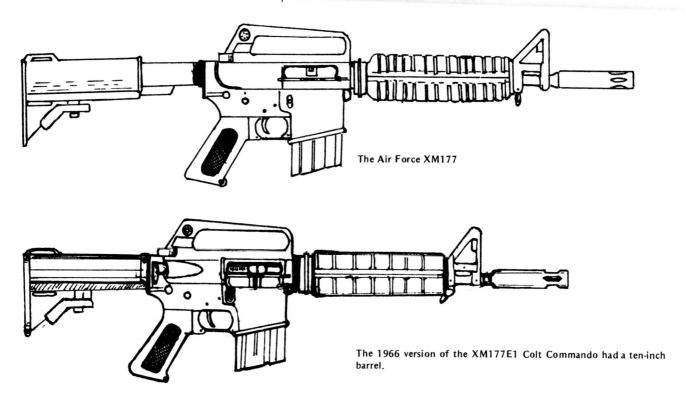

The Air Force XM177

The 1966 version of the XM177E1 Colt Commando had a ten-inch barrel.

Both the XM177 and XM177E1 had 10-inch barrels, and each had an empty weight of 5.5 pounds and a cyclic rate described somewhat evasively by Colt as being "650 to 850 rounds per minute," the speed differences being created by the type of ammunition used with the rifle.

The two Commandos had telescoping stocks which looked much like the fixed stock of the Colt survival rifle developed earlier for the Air Force. With the stock "folded" the rifle was 28.3 inches long; with the stock extended it was 31 inches long.

A year later, the Commando was modified slightly by giving it an 11-1/2 inch barrel that allowed the use of a grenade launcher that was being developed for the Army. This new version was redesignated the Colt Commando XM177E2 (the Air Force version was lengthened, too, by an inch, but still did not have the forward assist).

To add a bit of confusion to the "submachine gun" style of the AR-15, an experimental submachine gun which utilized 9mm Luger ammunition was also created. This was a true submachine gun which was operated by blowback—i.e., the chamber was never locked up—as are most modern submachine guns. The gun utilized a regular lower receiver and parts with only the barrel, magazine, and bolt being modified on the upper receiver for the weapon to work. Probably the only reason this weapon never caught on was that the Commando version of the CAR-15 worked just as well, used the same round as the CAR-15 rifle did, fired a more lethal round, and the 9mm was not then a standard round in the U.S. military stores.

With the trend toward adopting the 9mm as the U.S. standard pistol round—to conform to NATO

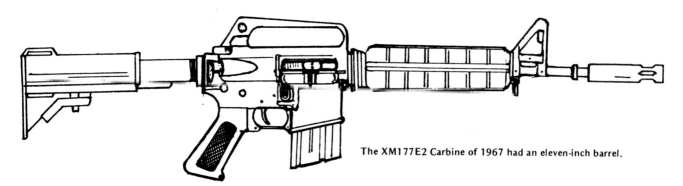

The XM177E2 Carbine of 1967 had an eleven-inch barrel.

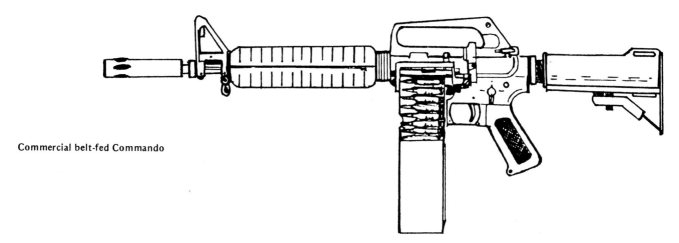

Commercial belt-fed Commando

ammunition use—Colt again introduced the 9mm Commando in 1984 for evaluation by the U.S. military and the law enforcement community.

In 1966, the Secretary of Defense decided that the AR-15 (XM16E1) would become the standard issue rifle for the U.S. military.

In its shortened Commando version, the Colt Automatic Rifle-15 (CAR-15) often made its appearance among the Special Forces or the CIA while the M14 was still being used by the regular soldiers. To the American GI, therefore, the shortened rifle was a "CAR-15" (Colt Automatic Rifle) while the regular-sized rifle (which was actually also a "CAR-15") was known by the military's designated numbers: "M16," "XM16E1," or "M16A1." This mixup continues today: the "CAR-15" is generally accepted as the shortened (Commando) version of the rifle, while the regular rifle is the M16. Further confusion has also been created by Colt's semiautomatic Sporter AR-15, which is the civilian version of the rifle and is aimed at the sporting market.

Additional confusion is caused by those who call the shortened version the "XM" since several non-Colt manufacturers are making rifles for sale to the civilian market and are calling such rifles XMs regardless of the barrel and stock size.

THE STONER SYSTEM

During the 1960s, the Stoner 63 Weapons System was competing with Colt for acceptance as the standard rifle of the U.S. military. It is interesting to note that three weapons—the AR-15, AR-18, and the Stoner 63 (which Eugene Stoner had had major influences in designing)—were all in the competition for acceptance as military rifles!

Eugene Stoner worked for Colt after leaving Armalite. After leaving Colt, he worked with the Cadillac Gage Corporation developing a family of weapons (perhaps an idea he'd picked up from Colt's extensive modification of the AR-15). At Cadillac, he secured his two co-workers that had been with him at Armalite: Robert Fremont and James Sullivan. Here they created the Stoner 62 system for the 7.62mm NATO round and the Stoner 63 family of weapons chambered for the 5.56mm cartridge.

The Stoner System consisted of fifteen assemblies which could be interchanged to create a number of configurations of weapons to suit any small

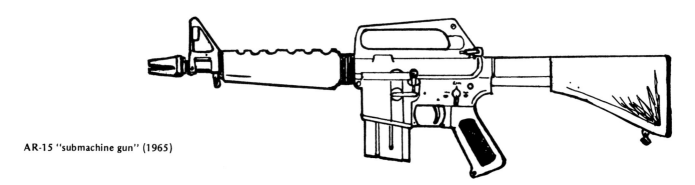

AR-15 "submachine gun" (1965)

arms purpose which the military might need while maintaining the use of the same basic control layout on each weapon and a large number of interchangeable parts. The first rifle that could be assembled was a short carbine with folding stock, then a full-size assault rifle, light machine gun (belt/magazine feed), a medium machine gun, and a tank machine gun. This system was evaluated by the Army for some time (as the XM22 Rifle, XM23 Carbine, and XM207 Machine Gun) and the Stoner light machine gun field-tested in Vietnam by the Marines (as the MK23 Belt-Fed Machine Gun). Only the MK23 saw much use in Vietnam where it was used by U.S. Navy SEAL teams.

Though the Stoner System had a lot going for it, the basic assault rifle was excessively heavy (7.75 pounds unloaded) since it had to have a rugged receiver for its use as a light machine gun.

Nevertheless, the idea of a family of firearms has caught on and has been successfully marketed by several companies, including Heckler & Koch.

THE COLT SYSTEMS RIFLES

Colt continued to develop new variations of the AR-15 to appeal to the military markets of the world. This soon developed into a case of "Systems Rifle Fever," and the late Sixties saw the development of other versions of the AR-15 by Colt. Like the Stoner "family" of weapons, the Colt system would offer the military the advantage of having several types of weapons which could easily be operated by troops familiar with just one of the weapons since the other firearms in the group would have common controls and the same general handling characteristics. Like other systems rifles, many parts were interchangeable, and all the weapons in the family used the same type of ammuni-

tion rather than the two, three, or even four different calibers of ammunition often encountered in a military's stores of weapons. These points were all definite pluses both logistically as well as in the field, where troops could exchange parts or modify their weapons according to their needs.

So, in addition to the Commando (XM177E1-E2) and the assault rifle AR-15 (M16) versions, the following weapons in this systems group were also developed:

1) The Heavy Barrel Automatic Rifle (HBAR-M1) was fed from the regular magazines. It weighed 7.6 pounds (empty) and was designed for use as a squad heavy automatic rifle (though it fired from a closed bolt). The HBAR had a cyclic rate of 650 to 850 rounds per minute.

2) The Heavy Assault Rifle M2 weighed 8.3 pounds (empty) and could be fed from regular 20- and 30-round magazines or a disintegrating belt. The M2 was also designed for use as a squad heavy automatic rifle. The M2 had a cyclic rate of 800 to 850 rounds per minute.

3) The Colt Machine Gun (CMG) weighed 11.5 pounds with its light barrel or 12.5 pounds with a heavy barrel. It fired from an open bolt (unlike the other rifles in the series) at 650 rounds per minute (650 to 850 rounds per minute with solenoid operation).

Though the CMG fired the 5.56mm rather than a heavier caliber, the machine gun was designed for use as a medium or light machine gun and had mounts for tripod, bipod, vehicle, or—with solenoid operation—helicopter/plane use.

The CMG series of machine guns must have appeared somewhat overrated in regard to the potential of the 5.56mm round for long-range use in a medium machine gun mode. The round was

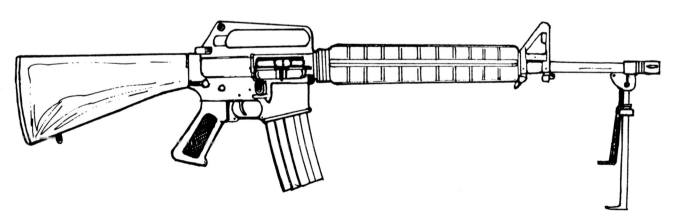

A version of the Heavy Barrel Assault Rifle is shown here.

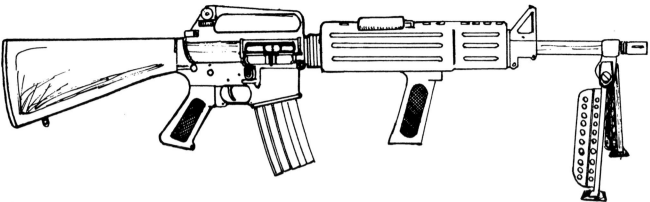

AR-15 Heavy Barrel Assault Rifle

upgraded during the mid-Sixties by Colt and the Frankford Arsenal with an experimental 68-grain bullet which—coupled with the 1-in-9 twist they developed for it—probably worked nearly as well as the 7.62mm NATO round.

The bullet itself had a history dating back to 1954, when a group of heavy .22-caliber bullets were made by Sierra for experimenters at the Aberdeen Proving Ground. Colt started with this 68-grain bullet and changed the ogive from a 7-caliber tangent radius (like that of the old military .30-

'06 M1 bullet) to a 10-caliber secant radius (like that of the 7.62mm NATO bullet). Just as the 7.62mm ballistics were improved over the .30-'06 by the change, so, too, was the new 68-grain bullet's performance improved over the older version originally made for the Aberdeen experiments.

These new bullets were manufactured and loaded into rounds by the Federal Cartridge Corporation during 1965 and 1966 for Colt to use in their machine guns.

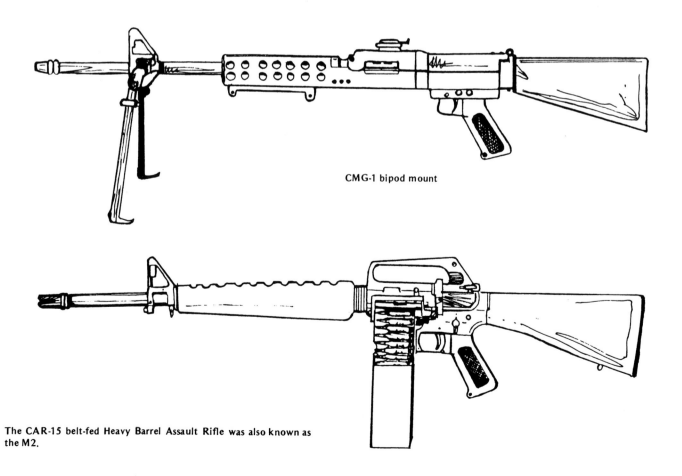

CMG-1 bipod mount

The CAR-15 belt-fed Heavy Barrel Assault Rifle was also known as the M2.

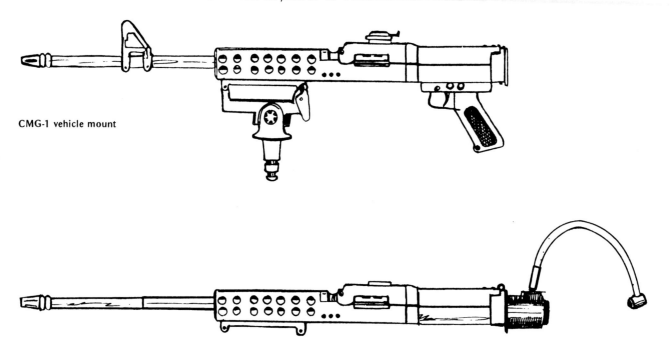

CMG-1 vehicle mount

CMG-1 selenoid operation

The round was later modified by the Frankford Arsenal and designated the XM287. It was used for a time in the Stoner 63 system when the military was evaluating it for possible use. The XM287 was manufactured for the tests by Valcartier Industries (in Canada).

The XM287 had a muzzle velocity of 2,960 feet per second which—coupled with its weight—gave it the ability to perform as well as the 7.62mm NATO round when it came to flatness of trajectory, accuracy, penetration, and ability to resist wind deflection.

Of all the services, the Marines apparently were the most interested in adopting the 68-grain bullet—and possibly some of the Colt machine guns—but did not do so because such a change would have required replacing the barrels on all of the 50,000 M16 rifles in their inventory. No decision to change was made by any of the other military branches, either.

So, even though the Colt family of weapons did offer many advantages, the weapons family was never accepted for actual combat use except for the Commando carbine and the AR-15 rifle (that was already in use as the M16 rifle). The HBAR found limited use in some squads.

The HBARs that saw combat were used as automatic rifles, with two HBARs issued to a squad. In theory the other squad members were to use the M16 in the semiauto mode only, while the HBAR was to be fired in the automatic mode. Though this idea was good on paper, convincing some of the squad not to ever fire in automatic mode was not too successful.

One development of the systems work quickly found its way to the battlefield: the 30-round magazine. The AR-15 was first issued with a 20-round magazine (because the original specifications given by the military called for a 20-round magazine), but the family system rifles demanded 30-round magazines for the increased firepower such weapons required.

At this point, the Stoner system used a 30-round magazine. The Viet Cong were armed with AK-47 assault rifles, which boasted a 30-round magazine. It was only a matter of time, therefore, before the 30-round magazine was introduced to U.S. troops in Vietnam.

The larger capacity magazine made its way into the hands of the troops on the Commando and the HBAR and eventually filtered down to those using the M16. Special LC-1 web gear was finally designed to hold three 30-round magazines, and the gear became standard issue.

PROBLEMS WITH THE AR-15

In 1967 the U.S. government purchased the rights to manufacture the AR-15 from the Colt Firearms Corporation. The government then entered into contracts with General Motors and the Harrington and Richardson Company to make approximately 250,000 rifles each for the military. Many believe that some of the General Motors guns had problems with quality control and performed poorly in Vietnam, thus greatly injuring the AR-15's reputation, which was already clouded by poor magazines, the use of ball powder in cartridges for the weapon, and the military's tinkering with the design of the weapon.

Soon, tales of soldiers killed in battle because their weapons had failed were making their way back to the United States. Congress launched an investigation, and the error of using ball powder in the weapons was brought to light.

As early as 1966, there was at least one report from the Army's CEDEC field testing group at Fort Ord which warned officials that though the AR-15 was much more effective than either the M14 or AK-47 (which it was being tested against), the problems produced by using ball powder were making the weapon jam to the point where it was unreliable. Unfortunately the Army ignored the warnings of those engaged in testing.

The public outcry was too great for the jamming problems of the AR-15 to be ignored, and the military made a number of changes to the rifle to help alleviate the problem:

- The weapons were chromed on the inside of the chamber, barrel, and bolt carrier. (Some bolts and bolt carriers were entirely chromed, but such chroming was found to cause excessive wear.)
- The military switched to an improved powder (though still not the IMR).
- A new stock was designed which had a cleaning kit.
- A new buffer was designed to slow down the rate of automatic fire.
- The soldiers were taught how to carefully clean and lubricate the rifles. (The military had been convinced the rifle did not need to be cleaned often and—with the humid Vietnam environment—problems of weapon fouling were greatly compounded by the lack of cleaning. Most rifles had even been issued without cleaning kits so that—even after the problems with the rifle were traced to not keeping it clean—there weren't enough cleaning kits to be issued to the troops for some time.)

The changes made in the rifle and the difficulty in keeping it clean cleared up the problems with the weapon, though it took many years for the weapon's tarnished image to regain its luster.

The new-style rifle was very similar to the XM16E1 but had the addition of chromed parts in it and a number of other minor modifications. Early in 1967 this rifle became the M16A1, and was finally recognized as the standard weapon of the U.S. military.

Late in 1967 it was discovered that not all of the problems with ball powder were the fault of "ball powder," per se. A study done by the Frankford Arsenal showed that both ball powder and IMR powders left residues in the gas tube and action of the AR-15 after extensive firings of 4,000 to 6,000 rounds. However, the ball powder residues tended to clog the gas tube so that not enough gas moved through it to function the rifle.

Continued investigation of the problem by Frankford Arsenal and Winchester-Western Powder Company found that the residue was comprised of calcium carbonate mixed with primer residue and bullet jacket fragments.

From here the trail led to the main problem that the ball powder was causing: too much calcium carbonate (used in the manufacture of the powder) was being placed in the powder. (Some of the ball powder was being remanufactured from World War II-Korean-vintage artillery powder, and excessive amounts of calcium carbonate were probably used when neutralizing the acid created by the deterioration of the powder.)

Thus, in 1968, amounts of calcium carbonate added to powder were lowered by 50 to 75 percent to prevent excessive fouling of the gas tube.

GRENADE LAUNCHERS AND THE AR-15 CARBINE

During this time Colt was working on the M148 grenade launcher, which was to be attached to the underside of an M16. This 2.8-pound accessory would allow a rifleman to have the punch of the single-fire M79 grenade launcher without sacrificing the use of his rifle. In fact, some of the troops in the field had already improvised such a weapon by cutting the stock off the M79 and cobbling it to the handguard of the M16.

In order to place the M148 grenade launcher on the short Commando, it was necessary to create a slightly longer barrel on the rifle. The barrel was then increased to 11.5 inches, and the suppressor was modified to allow the M148 to be attached to it.

Cross section of a 40mm cartridge, HE round

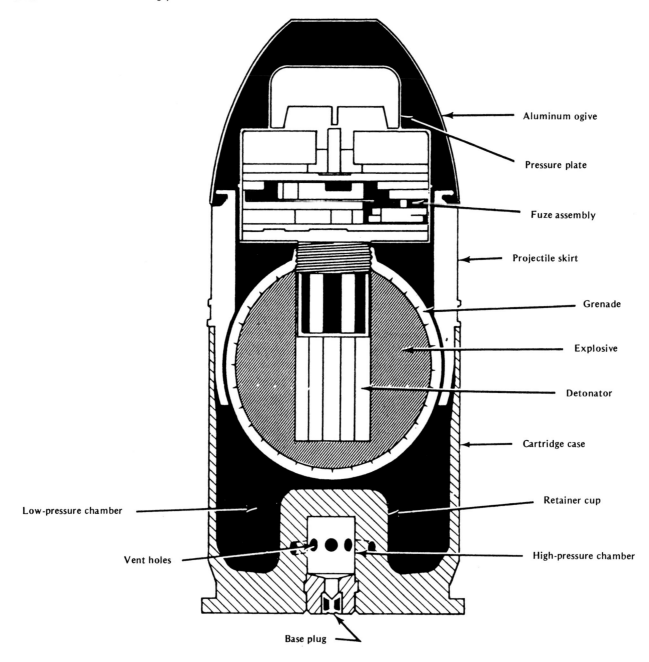

Aluminum ogive

Pressure plate

Fuze assembly

Projectile skirt

Grenade

Explosive

Detonator

Cartridge case

Retainer cup

Low-pressure chamber

High-pressure chamber

Vent holes

Base plug

Though the XM148 was a very useful weapon for troops in Vietnam, it suffered from problems which made it overly complicated and sometimes unreliable. Thus in 1967 the U.S. Army started the GLAD (Grenade Launcher Attachment Development) project.

Three companies worked with the Army to realize the designs produced by GLAD. They were the AAI Corporation, Aero-Jet General, and Ford Aerospace. AAI produced the best launcher, the XM203, which was standardized and accepted by the Army as the M203 in mid-1969.

Though early XM203s had a short barrel so that they could be mounted on the short Commando carbines, the final version of the M203 had a longer barrel to increase the range of the grenade launcher. Increased range was felt to be of more importance than the ability of the M203 to be mounted on both rifle styles.

Beginning in 1971, Colt became the sole producer of the M203.

A short-barreled AR-15 carbine is occasionally

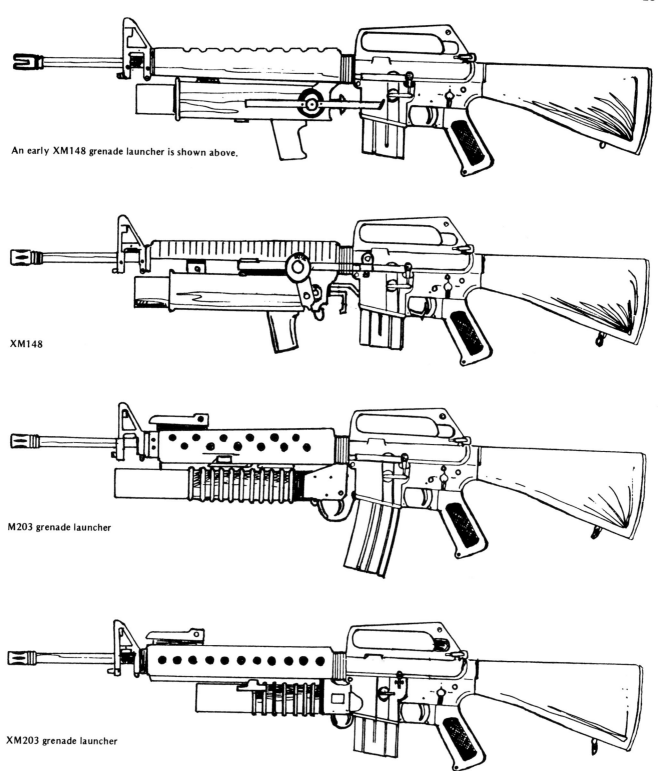

An early XM148 grenade launcher is shown above.

XM148

M203 grenade launcher

XM203 grenade launcher

seen in pictures and sales literature put out during this period. It is basically a full-sized rifle with the barrel cut down to 15 inches so that the flash suppressor is just ahead of the front sight assembly. Though a few of these saw limited use in Vietnam, the rifle was a poor second to the regular rifle or "submachine gun" Commando.

SMALL ARMS WEAPON STUDY

Starting in 1966 with the SAWS (Small Arms Weapon Study), the Army began working with the idea of producing a lightweight machine gun which could be carried at the squad level and which would use the same ammunition as the standard rifle. The machine gun had to weigh under 22 pounds with 200 rounds of ammunition and be effective beyond 800 yards. These two requirements were in direct conflict with each other since a machine gun that weighed 22 pounds *with* 200 rounds would have to fire the 5.56mm ammunition which wasn't effective out to that range! (Many authorities feel that the range requirement was put in by those who favored the 7.62mm NATO. The need for that long of a range flew in the face of earlier research that showed that most rifle casualties were produced within 300 yards—with the majority being under 100 yards.) The military wanted machine guns which could reach out to 800 yards, while the 5.56mm rifle round only was effective within 500 yards.

It was finally decided to upgrade the ammunition used in the rifle so that it would have the desired range while creating a minimum of change with the standard weapon in the U.S. arsenal.

Strangely enough, the Stoner 63 and Colt's CMG-1/CMG-2 machine guns all fell through the cracks of the SAW program because of the Army's change in requirements. Probably much of the reason that the U.S. military never used either of these weapons had to do with the over-800-yard-range conflict in the early specifications for the SAW.

The CMG-1 (Colt Machine Gun, Version 1) was started early in the Sixties as part of the CAR-15 Weapons System. Colt engineers, headed by Robert E. Roy, worked toward developing a lightweight machine gun which used as many of the AR-15 parts as possible to enhance the ability of the weapon to use parts already in the military's inventory. If the CMG was very similar to the AR-15, less training of soldiers would be required since they were familiar with the AR-15.

The CMG-1 was dropped in the late Sixties in favor of a new rifle design which would not necessarily use parts or design similar to the AR-15. This second version was designated the CMG-2 and developed by George F. Curtis and Henry J. Tatro.

The CMG-2 was tested during 1969 but was rejected because it did not have a high enough rate of fire and because of the SAW 800-plus yards requirement. Thus, the CMG-2 and Stoner 63 system machine gun have both been waiting in the sidelines, ready for use, since the late Sixties.

The Stoner 63 system was turned down in testing during 1968 through 1972 and the Colt system in late 1969. During the early part of 1972, the Army went to American manufacturers to have them develop candidates for the new SAW (Squad Automatic Weapon, Light Machine Gun).

The military originally hired only two companies: the Maremont Corporation (with the XM233) and Ford Aerospace (with the XM234). No attempt was made to try the Stoner or Colt machine guns which had already been developed!

Soon a third machine gun was added when Rodman Laboratory added their version (the XM235) to the contests, but Stoner and Colt Firearms never submitted their weapons to the tests.

In the meantime, the U.S. Marines decided to produce their own "interim SAW" rather than wait for the results from the long, drawn-out tests. For some time, Colt had been working with a modified heavy-barreled version—the HBAR—of the AR-15 which could be used as a SAW. (One version, the M2, even had a belt/magazine feed capability.)

Rather than use the Colt rifles that had been developed, the U.S. Marines decided to try a version of their own and experimented with a heavy barreled M16. It had been modified by WAK, In-

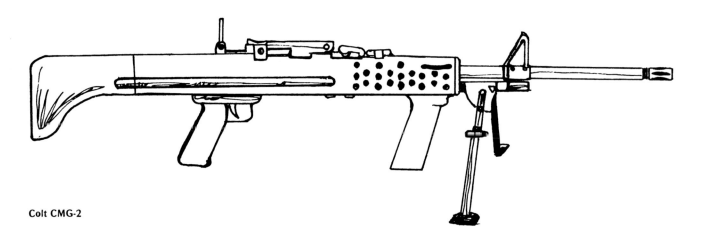

Colt CMG-2

corporated, with a special buffer (which allowed the machine gun to fire from an open bolt) and a new muzzle compensator to prevent muzzle climb with the lightweight gun. This project didn't work out well, however, and was abandoned in 1977.

The military's work on the SAW program continued. The XM235 was modified to become the XM248, and three more weapons were submitted by other companies to the trials: the FN Minimi (XM249), Heckler & Koch 23 (XM262), and a heavy-barreled AR-15 machine gun (XM106).

THE XM106

Work on the XM106 was started in 1978 by the Ballistic Research Laboratory (BRL) at the Aberdeen Proving Ground.

This modified AR-15 rifle was very similar to the Army's standard M16A1 but had half the front grip mounted permanently in place, while a quick-change, detachable barrel unit was mounted into it. (Some versions also had the front sight base moved forward on the barrel to create a longer sighting radius.) A standard M2 bipod (used with the M14 rifle) was added to the center of the handguards with a pistol grip mounted under the bar-

rel, along the lines of the Thompson submachine gun. (The bipod was probably located toward the receiver to prevent the barrel being flexed when the bipod was not in use. Standard bipods on the M16 cause a slight change in the point of impact of the bullets.) The rear sight was originally a modified M60 machine gun sight but was replaced with a leaf sight which had settings for 300, 500, 800, and 1,000 meters.

This weapon was designated the XM106 during the trials and could use regular 20- or 30-round magazines, an 83-round drum which looked much like the drum used with the Thompson submachine gun, or special "Tri-Mag" assemblies. (The Tri-Mag was developed as a part of the WAK project of 1977.) The Tri-Mag was basically three 30-round magazines joined together. With a cyclic rate of 750 rounds per minute and a weight of 10.6 pounds (empty, with bipod), it was one of the lightest weapons in the trials.

But the XM106 did not make the grade; the Fabrique Nationale M249—most reliable of all the weapons tested—was chosen as the SAW in 1980.

Throughout the tests the heavy-barreled AR-15 was used as the yardstick against which other

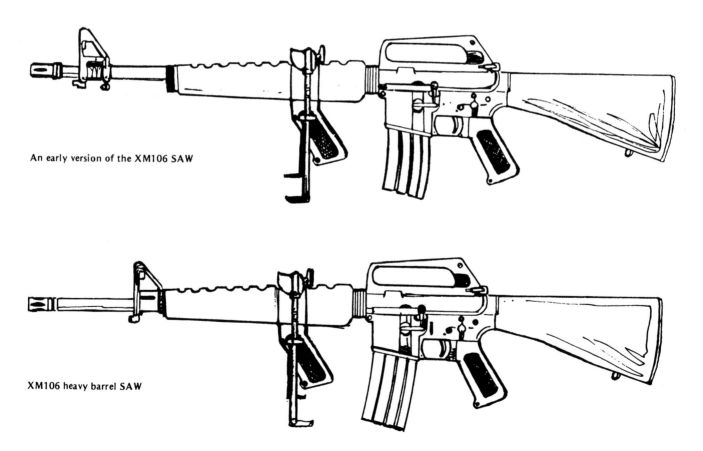

An early version of the XM106 SAW

XM106 heavy barrel SAW

machine guns were measured. One can't help but wonder if there weren't some military decision-makers who wanted the heavy-barreled AR-15 to win the trials, while others were doing all they might to make it lose.

In the end, one of the deciding factors was the XM106's inability to feed ammunition from a belt. Since all the entrants were required to have feed wells for standard magazines (apparently so other members of the rifle squad could "feed" the machine gun with the magazines they carried), one has to wonder if the need for a belt feed was necessary.

Another problem was the excessive modification necessary to make a quick change barrel. Since it is very easy to remove the whole upper receiver, bolt carrier, and barrel assembly on the AR-15, some critics argue that it would make more sense to use a regular receiver/barrel setup and change the barrel *and* the receiver. This would be quick and easy to do, and the barrel could be sighted in beforehand (unlike any of the other quick-barrel-change systems).

Using this line of argument, several heavy barrels could be issued to the squad and used with three-round burst operation (which the SAW project showed was much more effective than long bursts of fire). An added benefit would be that any barrel that did overheat could be exchanged with any of the other receiver/barrel assemblies of the squad. The weight of such a system would still come in at less than that of the 14.26 pound FN Minimi: a heavy, 24-inch barreled AR-15 weighs only 10.5 pounds. The extra 3-plus pounds saved translates into 90 rounds of extra ammunition packed in magazines.

Though the USSR is not the best of examples, it is interesting to note that its counterpart to the U.S. SAW is the RPKS74, which has a 40-round magazine or 75-round drum magazine. The RPKS74 fires from a closed bolt and operates in the same manner as the standard rifle. Little time is lost, therefore, in training troops to operate the weapon.

THE M16A1

The current versions of the M16A1 Commando-style rifle offered by Colt for military use are created by combining the M16A1 action with a telescoping stock and a 14.5-inch barrel. It is mounted with the regular flash suppressor used on the M16A1, and enjoys brisk sales worldwide even though the U.S. military's interest in it appears to have waned. The current short rifle is generally known as the "M16A1 Carbine." A few M16A1 Carbines are also being made by Colt with the same short 14.5-inch barrel. These weapons have a standard AR-15 rifle stock rather than the telescoping stock.

This Commando-style weapon (again mistakenly called the "CAR-15" by many users) seems to have struck the perfect balance. It retains its handiness, while having a barrel long enough to maintain a large part of the round's velocity potential, keeping the flash and noise created by unburnt powder to a minimum. (There is still, however, a small potential for jamming since unburnt powder makes its way down the short gas tube.) It should be noted that the current version of the short AR-15 also sports a bayonet lug unlike the former Commando versions or the Air Force Survival Rifle.

Though the Commando is not officially included in the U.S. military inventory, it was to be seen in the 1983 invasion of Grenada and is still in use by the Air Force Security units, Special Forces, and the Rangers.

COLT'S AR-15 SPORTER RIFLE AND M231

Domestically, the Colt Firearms Corporation

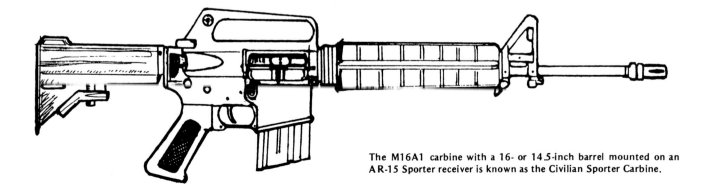

The M16A1 carbine with a 16- or 14.5-inch barrel mounted on an AR-15 Sporter receiver is known as the Civilian Sporter Carbine.

offers the "AR-15 Sporter Rifle" which is a semi-auto AR-15 without the bolt assist. The upper and lower receiver are modified somewhat to discourage the use of M16 receivers in conjunction with an upper or lower AR-15 receiver. Internally the hammer, trigger, disconnect, selector, and bolt carrier have all been changed to discourage any tampering that might convert the weapon to an automatic mode. (Though all parts mentioned can easily be replaced by M16 parts, the automatic sear from an M16 cannot be placed in the weapon since a hole for the pivot pin isn't drilled in the civilian AR-15.) The bolt carrier beneath the firing pin has also been milled out so that the commercial hammer will prevent the gun from functioning if the disconnector is removed in an effort to create a "slam fire" auto weapon.

The commercial Sporter Rifle was originally marketed under the name "Colt Commanche." This name was later dropped because a commercial aircraft had been marketed as the "Comanche" just before Colt brought out its rifle.

An "AR-15 Sporter Carbine" is also available from Colt to citizens of the United States. It has the telescoping stock of the Commando and a 16-inch barrel with a regular AR-15 style flash suppressor. It is identical to the domestic AR-15 Sporter Rifle except for the shorter barrel assembly and the telescoping stock.

A recent entry into the military arms arena from Colt is the M231 port firing weapon. It makes use of many AR-15 parts and is designed to be fired by troops in armored personnel carriers.

Work was started on the modified AR-15 in 1972 when an FPW (Firing Port Weapon) was needed for the XM723 MICV (Mechanized Infantry Combat Vehicle). Much of the work on this was handled by the Small Arms Systems Laboratory at the Rock Island Army Arsenal. Two other weapons were tested with the modified AR-15: the M3A1 submachine gun ("Grease Gun" in .45 ACP) and the Heckler & Koch 33 (in 5.56mm).

THE XM231

The modified AR-15 proved to be the favored

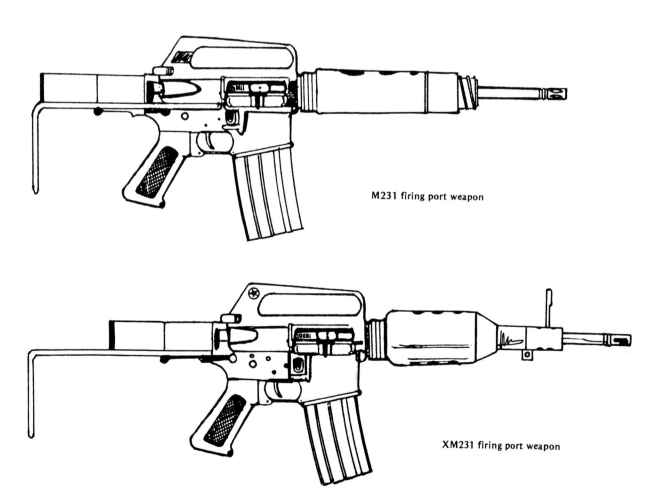

M231 firing port weapon

XM231 firing port weapon

choice for the FPW, so it was designated as the XM231 in 1974, and the decision was made to further develop it.

Firing from an open bolt, the original XM231 had an extremely high cyclic rate of 1,050 cycles per minute. Working with the FPW, the developers found that the cyclic rate needed to be dropped to 200 cycles per minute to obtain the best probability of making hits on a target. (The weapon was aimed via tracer rounds since sights could not be used because of the port/viewing plate setup.)

By 1976, all tracer rounds were being used in the XM231, and a telescoping wire stock and flip-up front sights had been added to the weapon so that it could be removed from the port and used by soldiers in the armored personnel carrier when they left the vehicle.

Many of those in authority felt that the sights and stock were unwarranted since the troops in the MICV were to have their own rifles to carry out when they left the vehicle. In a strange sort of compromise, the final version (completed at the end of 1979) has the stock but no sights! (It does retain the scope mount hole in the handle so that an enterprising trooper could fix a scope on the weapon.) Additionally, the cyclic rate in the final version of the M231 was raised from the accurate 200 rounds per minute back to the less accurate 1,100 to 1,200 rounds per minute.

The parts of an M231 are interchangeable with some 65 percent of the standard AR-15 parts. Some design features are quite different to allow it to meet the requirements set for it.

The recoil spring/buffer system has been shortened through the use of three nestled springs, and a large screw mount is located at the front of the handguard to allow soldiers to actually screw the rifle into its firing port. The gas tube on the M231 has been shortened and the weapon modified to vent its cycling gases outside the personnel carrier.

The M231 fires from an open bolt, so it has a heavier, floating firing pin with a striker which moves inside the rear of the bolt carrier. This striker follows the firing pin as the bolt locks shut, and the striker's momentum fires the cartridge. This does away with the need for a hammer in the M231. Other differences between the M231 and the AR-15 include: a modified trigger sear which holds the bolt in an open position until the trigger is pulled; an empty cartridge collection bag; and an extra heavy barrel to prevent overheating from long bursts of its all-tracer fire.

The M231 will weigh 8.5 pounds (empty) and is 28.5 inches long with the stock retracted (33 inches long with the stock extended).

THE "ARM GUN"

In the late Sixties, Colt started work on another interesting weapon which may prove to be the stepping stone toward future small arms development in the United States, especially in the area of submachine gun and pistol replacements. This weapon made use of many of the AR-15 parts but had the pistol grip mounted in front of the magazine. Unlike the standard "bullpup," this gun didn't have a stock.

This configuration was the "arm gun" which was described by Colt as a "lightweight rifle/submachine gun" for lack of a better term.

The stockless arm gun was aimed by holding the weapon straight out in front of the firer, with the shooter's free hand holding the butt of the weapon against his upper arm. The user then aimed by resting his head against the upper part of the arm holding the weapon and sighted down a small scope or the movable sight (which swiveled to accommodate right- or left-handed users).

The arm gun was originally chambered for pistol ammunition, but the recoil seemed so mild that the 5.5-pound weapon was chambered for rifle ammunition. According to users, the recoil even with 7.62mm NATO ammunition was minimal because the hand and arm soaked up the recoil rather than the shoulder as is the case with regular rifles.

In 1969 the Air Force tested the arm rifle for possible use as a survival weapon or a 5.56mm "submachine gun." The arm rifle used in the tests was chambered for the .221 IMP cartridge (so that now the weapon is sometimes called the "IMP").

As of this date no other real interest has been shown by the U.S. military in Colt's arm rifle. If the U.S. military should ever decide to change over to a bullpup-style rifle, it is possible that the arm gun design may be given a short stock and a bit longer barrel and offered for testing.

THE BUSHMASTER PISTOL

A slightly modified version of the arm rifle did come into production via the commercial market. Known as the Bushmaster "pistol," it was designed by Mack Gwinn. This weapon uses many of the design features of the Colt's arm rifle but, in fact,

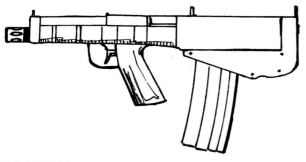

Colt ARM Rifle

Bushmaster pistol

uses even more of the original AR-15 parts to keep the costs of manufacturing the firearm down. A large number of spare parts are available on the military surplus scene at very reasonable prices, thereby keeping production costs down for a small manufacturer.

Because the rifle is used without a conventional stock, the rifle is considered to be a pistol by the BATF. Thus, it is one of the few "rifles" which can have a barrel of less than 16 inches in accordance with BATF regulations.

Early Bushmasters suffered from poor quality control, and the original manufacturer soon went bankrupt. A new company has been set up and—hopefully—the quality of the firearms being produced will be better than those first marketed.

A second model of the Bushmaster has also become available, bringing the weapon full circle: A stock was added, along with a longer barrel, and the pistol grip was moved back into a conventional position. The Bushmaster rifle, very similar to the AR-15 (and using a large number of AR-15 parts) is—for all purposes—really a stamped steel version of the AR-15 with a piston/recoil spring assembly over the barrel rather than a gas tube.

The Bushmaster pistol weighs 5.25 pounds and is 20.5 inches long with a 11.5 inch barrel. The rifle is 7.5 pounds (empty) and 37.5 inches long (with the wooden stock or when the folding stock

is extended) or 27.5 inches long with a closed folding stock. The barrel on both versions of the Bushmaster is welded to the steel receiver and has a twist of 1 turn to 10 inches.

Currently the Bushmaster is sworn by, or sworn at, depending on the quality control background of the rifle. It is hoped the continued improvement of quality control will keep this well designed rifle in the marketplace. (Rifles/pistols made before the quality control improvements *did not* have a "J" prefix on the pistols' serial number or an "F" prefix on the rifle. If you purchase one of these weapons, try it out and look for the "J" or "F" prefix.)

FOREIGN MANUFACTURE OF THE AR-15

Colt has manufactured a number of AR-15s for export to foreign countries, and several countries have also been licensed by Colt to manufacture their own AR-15s. The countries who manufacture the AR-15 include Singapore, the Philippines, and South Korea. The export and foreign manufacture of the rifles, plus the large stocks that were lost during the downfall of South Vietnam, have made the AR-15 one of the most used rifles in the world—rivaled only by the much older AK-47.

Taiwan has produced its own version of the AR-15 which its military has adopted as the Type 68 Rifle. This version has a lowered front

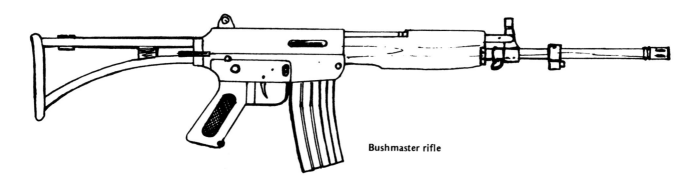

Bushmaster rifle

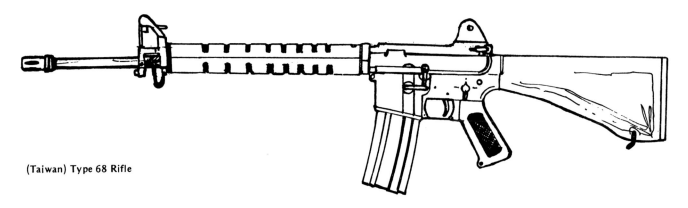

(Taiwan) Type 68 Rifle

sight, altered swivels, a lowered rear sight, reshaped stock, and redesigned handguards. The Taiwan design also did away with the carrying handle. The rifle weighs 7 pounds (empty) and is 39 inches long.

NATO countries have exerted pressure to change the AR-15 through the development of long-range 5.56mm ammunition. The SS109 5.56mm round was chosen to replace the old round used by the U.S. military. This round will be designated the "M-855" in the United States; its bullet weighs 62 grains (as compared to the old 5.56mm bullet of 55 grains). The new bullet has a solid-point/boat-tail design and is nearly as effective in many ways as the 7.62mm NATO round. Effective combat ranges will go out to 1,000-plus yards.

NEED FOR A FASTER TWIST RATE

The new 62-grain bullet will require a much faster twist rate to maintain its stability and accuracy. (Though it can still be used in the M16A1 with its 1-in-12 twist, its accuracy is poor.) Currently a 1-in-7 twist is being planned. Unfortunately the new twist rate is faster than is needed for the new M855 rounds in order to accommodate the longer M856 tracer round. (A 1-in-9 would give sufficient stability for the regular ball ammunition while maintaining a higher wound potential.)

The need for a faster twist rate dictated by the new ammunition meant that barrels of the current issue AR-15 of the U.S. military (M16 and M16A) would have to be changed or replaced to use the ammunition while maintaining accuracy. The Marines had aging AR-15s which needed to be replaced. The time was ripe to make some major changes in the M16/M16A1 rifle designs since the rifles would need extensive modification or replacement.

THE M16 PIP

During 1978 and 1979, the United States Armed Forces worked in conjunction with Colt to produce an M16 PIP (Product ImProved M16).

Many of the changes in the M16 PIP were in fact changes that Colt had been trying to make for some time, but the U.S. military had been rejecting. At any rate, when the testing was done and the results weighed, the following changes were found to be desirable in the AR-15 used by the military:

- The front sight would be a square stationary post.
- The windage and elevation adjustments to the sighting system would be made on the rear sight with a dial that can easily be operated with the fingers (similar to the one originally on the AR-10).

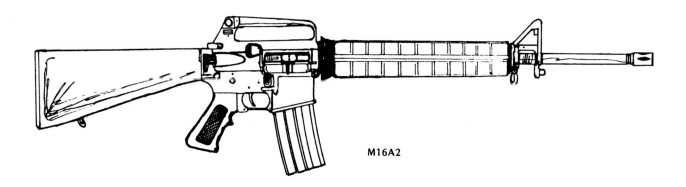

M16A2

- A large ring rear sight aperture was added.
- A new muzzle-brake/flash suppressor would be used to reduce the upward movement of the barrel during firing of bursts (similar to the M16A1 suppressor, but the lower "cut-outs" were missing). Originally a washer was to be used to cant the slits slightly to the right or left to compensate for left- or right-handed users.
- The barrel was made heavier from the front sight forward (with a 1-in-7 twist as mentioned earlier).
- Handguards were of the round-ribbed style originally on the Colt AR-15 with the two halves interchangeable.
- The slip ring was tapered.
- The end of the stock was lined to reduce its tendency to slip off the shoulder during firing.
- The stock became five-eighths of an inch longer to increase the ease in holding the rifle in the firing position. Old stocks were retained for use by shorter soldiers.
- The stock was made of a new plastic material ten times stronger than the old fiberglass stocks.
- The pistol grip has a finger grip ridge.
- A brass deflector was added to the upper receiver, especially welcomed by many left-handed shooters.
- The rifle has a three-round burst selection rather than a full-auto option on the selector. It will thus still have three positions: "Safe," "Semiauto," "3-round Burst."

The new M16 PIP was designated the M16A2 (for a time it was called the "M16A1E1") and weighs 7.2 pounds empty and 8.2 pounds with a 30-round magazine.

Interestingly, one of the modifications which Colt had developed and originally placed on the experimental Colt Automatic Rifle 703 was not added to the M16A2. This modification was a gas system which used a rod similar to that of the FN FAL or AR-18. The system was designed to prevent the fouling problems created by dirty powder and the gas tube.

Military planners were at odds on whether or not to adopt the M16 PIP. The "old school" was still griping because the current battle rifle isn't chambered for the old 7.62mm NATO cartridge, though undoubtedly tests with the new M855 round changed a few minds since it performed better than the 7.62mm NATO. The "forward lookers" wanted to wait until more radical weapons using caseless ammunition or weapons which conform to the CAWS (Close Assault Weapon System) were available.

At any rate, the U.S. Marine Corps *had* to have a weapon to replace the aging M16, so the Marines started adopting the M16A2 as their service rifle in the fiscal year of 1984. The Army wanted to wait to see if a caseless ammunition would be developed before it decided to modify the M16s in its inventory.

As it turned out, the German work with caseless ammunition had not progressed as quickly as had been hoped. The United States Army, too, is now carrying the new M16A2 version of the AR-15.

Many view the burst feature as a marked improvement over full-auto fire, though a good argument might have been made for making the guns fire only in the semiauto mode if conservation of ammunition is necessary. (This is the practice of Canadians, who limit the automatic fire mode to only several members of a squad.) Many wonder, too, how often a three-round burst would actually improve the kill ratio during combat conditions. Their argument is this: If one round of the 5.56 is really effective, then why are three needed? If three are needed, why ever use semiauto fire?

Supporters of the three-round burst control argue that the SAWS (Small Arms Weapons Systems) studies of 1967 to 1980 found that the three-round burst was greatly superior to long bursts of fire. (When a tape or film of troops who have automatic modes on firearms is studied, it is very common to see soldiers fire off a whole magazine of ammunition—often without aiming. Certainly the three-round burst control would cut down on such waste. But again, one might argue that the waste would also be stopped with better training or semiauto-only weapons.)

The three-round burst also has a quirk which shooters used to firing short bursts on the current M16A1 may find exasperating: A person can fire only one or two rounds when in the three-round burst position, and then the remainder of the three fire off the next time regardless of how long the trigger is held the second time.

The heavy barrel was added because the M16A1 barrel was sometimes bent. In fact, problems with bending barrels were limited to when the rifle was used with a bayonet on it and when troops misused the rifle for a pry bar to open crates. Since the last time a bayonet was used in modern warfare for anything other than guarding prisoners was during the Korean War (and even then entrenching tools and the like were often just as good), it seems doubtful that modifying the M16 for heavy bayonet use is going to improve the troops' ability to survive on the modern battlefield (more on bayonets in the accessory chapter of this book). Likewise, one can't help but wonder if the added cost and weight of the barrel are justified in creating a better pry bar that doubles as a rifle.

BULLETS

The 1-in-7 twist rate for the 5.56mm was originally developed in Europe, along with the new bullet, to create a more "humane" bullet. The theory is that the lighter 55-grain bullet which tumbled upon impact was more cruel than a heavier bullet that drilled and then tumbled after shattering bone or dense tissue! Worse yet, soldiers may start shooting their opponents a number of times if they feel one bullet is not effective. The bottom line will probably be that those shot with "humane" bullets will probably be in worse shape than those shot by "inhumane" bullets.

Humane ideals aside, there is some question as to whether or not the new, more stable round is more lethal. Preliminary tests seem to indicate that it is just as lethal—perhaps more so—than the 55-grain bullet. It does, however, have a greater range. Probably only combat will tell how effective the new round really is.

An added benefit for the short run is that the new M16A2 can use the older ammunition without a loss of accuracy, though the bullets are not as unstable when fired from the new barrel.

TRENDS

The AR-15/M-16 series of rifles have a colorful past during their short history, and it will be interesting to see what the next improvements and modifications will be on this rifle, so popular worldwide. With Colt's willingness to modify the AR-15 to suit the needs of the arms market, it appears very likely that many new modifications will be on the horizon. Likewise, the popularity of the rifle on the civilian and law enforcement fronts guarantees that a large number of aftermarket accessories and modifications will also be showing up to help owners of this rifle modify it to suit their needs.

Colt has also taken advantage of their retooling changes for the M16A2 by introducing a "Colt AR-15 A2 Sporter II" with the same barrel (1-in-7 twist) as the M16A2. The Sporter II also has a flash suppressor without lower vent cuts, square front sight (which must be moved for adjusting for bullet drop), and a pistol grip with the extra finger grips as well as grooves down the grip's back. Except for these changes, the Sporter II is identical to the Sporter I.

At the same time, there are trends that may bring a halt to military use of the AR-15.

The current trend in military rifles is to go with a "bullpup" design. The bullpup design is hardly new; General G. Patton had a bullpup modification of his own before World War II. It creates a light, short rifle without shortening the barrel to the point where it cuts down on the bullet's velocity.

The bullpup design places the pistol grip ahead of the rifle magazine and most of the mechanism of the receiver. The rear of the receiver becomes the stock. The rifle is shorter by the length of a regular stock.

The AR-15 cannot be adapted to the bullpup very easily. The buffer tube can be shortened considerably and the pistol grip replaced by using the magazine as an oversized grip so that the trigger and trigger guard are just ahead of the magazine. Though this creates a short, handy weapon (especially if a 10-inch barrel is used), the magazine is rather large for most shooters' hands and certainly is not the most ideal of pistol grips.

It would appear that the AR-15 will have to have its recoil/buffer action modified by either placing the recoil system over the barrel or replacing the bolt carrier with one similar to that of the AR-18, which has dual springs in the rear of the receiver. (The AR-18 would be easily adapted to the bullpup design.) Probably either of these changes would require a change in the upper receiver. Unless the rifle were equipped with an optical sight, the upper receiver would also have to be modified to place the rear sight forward of its current position. Finally, some sort of bar would be needed to connect the trigger to the receiver.

All in all, it would be easier to create a new rifle.

The rifle configuration is not the only point at which major changes may be occurring in battle weapons in the not-too-distant future. There are currently two U.S. military projects which may lead to the development of new weapons. One program is the ACR (Advanced Combat Rifle) and the other the CAWS (Close Assault Weapon Systems) program.

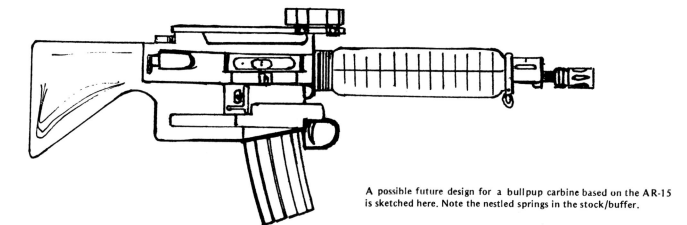

A possible future design for a bullpup carbine based on the AR-15 is sketched here. Note the nestled springs in the stock/buffer.

Late in 1982, the U.S. Army issued two contracts (one to the AAI Corporation and the other to Heckler & Koch, Incorporated, in the U.S.) for the development of caseless ammunition and interim rifles, as well as prototype rifles for testing and demonstrating the ammunition.

Caseless ammunition is hardly a new concept. The German military had actually produced some experimental caseless ammunition (and weapons for it) during World War II. In the Fifties, the U.S. Frankford Arsenal started experimenting with ideas for caseless ammunition, and in the Sixties actually developed a 7.62mm caseless round which had ballistics similar to the 7.62mm NATO round.

The Frankford Arsenal work continued into the Seventies, and the U.S. Army turned the work on it over to American contractors.

Caseless ammunition would offer a number of advantages to the design of a rifle. The rifle would be lighter and simpler since a mechanism to extract or eject fired cases would not be needed. The weapon would be more resistant to dirt and mud since an ejection port does not have to be part of the receiver since there are no empty cases to eject (though some sort of covered port is necessary for the removal of any misfires). Without a metal case, the ammunition would be lighter, too, which would allow the soldier to carry greater quantities of ammunition.

With the simplified mechanism of the rifle, the weapon can cycle very quickly (up to 2,000 rounds per minute). If it is fired in an automatic or burst mode, the recoil becomes a steady push rather than successive jumps, and barrel climb is nonexistent. This means that a very small caliber bullet might be used since it would be easy to create multiple hits. (Multiple wounds are more deadly than a single hit; the lethality increases by the square of the number

of hits. Three hits are nine times as bad as being wounded by a single round of the same caliber and velocity.)

But despite the advantages of caseless ammunition, early caseless ammunition had problems with staying together during rough handling and with cook-offs when the chamber became overheated. With conventional cartridges, the brass absorbs much of the heat created by the powder of the fired round.

New caseless ammunition produced by Dynamit Nobel for the Heckler & Koch rifle (the G11) can be handled quite roughly and is waterproof. Cook-off problems seem to be almost overcome in the current bullpup design of the HK G11. The rounds are small rectangular chunks of solid "powder" molded into a body around a bullet, which sticks out of the front of the "cartridge" of powder about as far as a bullet does from a regular round. The primer is located in the center of the tail end of the round. The current bullet is 4.7mm in diameter and the entire round is about half the length of a 5.56mm NATO round.

The AAI Corporation ACR follows current rifle design and has the ability to fire from either the closed bolt or open bolt—a definite plus when heat buildup may be a problem. The AAI ACR will use the major idea that came from the SALVO research: firing short bursts of low recoil ammunition at a high rate of fire.

The CAWS program seems to be aimed at producing an updated selective fire shotgun which would fire ammunition capable of being used at almost twice as great a range as is conventional shotgun ammunition. The ammunition itself will be held in detachable magazines to allow for quicker reloading of the CAWS.

Late in 1983 two CAWS contracts were given

to the AAI Corporation and Heckler & Koch, Incorporated. Ammunition is being produced for the CAWS by the Olin Corporation. Currently it appears that round will have a diameter of 18.4mm —equal to a 12-gauge shotgun round so that conventional shells can be fired from it as well as special shells made just for the CAWS.

The Heckler & Koch weapon is outwardly very similar to the AAI ACR: It makes extensive use of plastic furniture and uses a bullpup design. The AAI, of more conventional design, is similar to current sporting arms except for the detachable box magazine.

Although the CAWS certainly offer a lot of the pluses needed (according to the extensive studies done by the military since World War II), there are a lot of problems with the systems. One is the weight of the weapons and the rounds. The HK CAWS weigh 8.8 pounds empty and 10.9 pounds with a ten-round magazine in place. The ammunition is heavy and bulky (as anyone who has used a conventional shotgun in combat knows). The size of the ammunition also limits the number of rounds that can be held by a magazine to only ten. Though each cartridge contains a number of pellets or flechettes, there are only ten "bangs" before it's time to reload, and there is a limit as to how many loaded magazines can be carried due to their size and weight.

Critics of the CAWS project feel troops carrying a CAWS might be out of ammunition pretty quickly in battle—especially if the automatic-mode option is offered on the weapon.

With this weight problem, as well as its maximum range requirement of 150 to 200 yards, it would appear that the CAWS program might be headed the way of the SPIW program, which bogged down due to similar problems.

Another development which may soon enter the small arms arena is a new cooling system for the barrels of automatic weapons. The system, proposed by Stuart A. Hoenig of the University of Arizona, seems to pave the way for light-barreled weapons which could fire a large volume of small bullets from a closed bolt.

The system takes advantage of a well-known electrical phenomenon called electrostatic wind. Electrostatic wind is created when a high DC voltage is generated; the oxygen molecules are pushed away from the charge and collide with other molecules in the air, thereby creating a miniature windstorm.

Since the electrical field can be generated by permanent magnets or piezoelectric elements, batteries are not needed for the system to operate since it could easily be powered by the firearm's bolt movement or by gas bled from the barrel.

This electrical charge would be stored in a small capacitor, and the electronic breeze could then be directed down the outside of the barrel. The air movement would continue for some time after the weapon was fired.

The system would be very light, could be totally sealed against dust and mud, and would be extremely rugged. If the system should fail, the weapon would still function, though heat buildup would be a problem if the weapon were fired as it normally would be with the system functional. (The trend toward lighter weapons firing a number of small rounds at a time may just be beginning.)

In any case, it is pretty certain that the AR-15 family of weapons will be in use by the military at least to the turn of the century. If the AR-15 proves to be as durable as the Colt .45 auto pistol, the AR-15 may even see military use through much of the next century.

2. Magazines

One of the most important parts of the AR-15 is the magazine. It is also one of the parts that users often skimp on. Poor magazines will sooner or later guarantee failures in feeding—and may do so when your life is on the line.

The most sensitive parts of a metal magazine are the lips. Care must be taken not to drop magazines on their lips, or allow them to scrape or bump into other objects. Loaded magazines should be carried in pouches designed for them, not cloth military bandoliers or the like.

It will probably only be a matter of time until the U.S. military develops a flak jacket with outside pockets to replace the web gear. Until then, we'll have to make do with the sometimes awkward combinations needed to carry gear and magazines.

Probably the best way to carry spare magazines is in some of the newer SWAT-style vests that have come onto the market, since pockets are easily accessible. One of the best ones I've seen to date is available from Newman's GI Supply, RR No. 1, Box 782-LS, Augusta, NJ 07822, for $76. It has pockets to hold up to eight 30-round magazines, two first-aid/compass-sized loads, and maps or the like in two large inside pockets. Canteens, knives, etc., should probably be worn on a belt. The greatest problem with some vests on the market is that they try to put every bit of gear on the vest, which can be hard on one's back since the load is being carried there instead of partly on the waist and lower back. The Newman vest does not create this problem since it has limited the number and size of pockets.

Most other vests on the market are on the expensive side (though worth it). Many times cheaper military surplus pouches are very satisfactory and can be carried comfortably on the military web belt and suspenders. One pouch can easily be carried on a regular belt if you have a limited need for ammunition; this is not always wise, but sometimes more practical.

Though the military pouches are good in protecting and carrying magazines, they are not good at allowing you to remove the magazines in a hurry. The best solution to this is to practice a lot at getting them out. In combat, don't put the empty magazines back into the pouch since this will create too much confusion as to which magazines are empty. Carry a large pouch just for empty magazines or place them by you so that they can be quickly gathered and placed in a pouch when all three magazines are empty.

A "Six-Pack" pouch is available commercially for carrying 30-round magazines. It has three pockets on each side and is carried by a strap over the shoulder. One real plus is that it can quickly be plucked from a closet or car trunk along with your rifle. You're then ready to go without a lot of belt fastening and gear adjustment. The tough olive green Six-Pack is made of ballistic nylon so that it wears about forever; it's available from Sierra Supply (Box 1390, Durango, CO 81301, 303/259-1822) for $19.95.

To make the removal of magazines quicker, some users put a tape tab on the magazine so that it can be pulled out by the tab. Just be sure it is really secure on the magazine. In pouches that hold groups of two or three magazines, you need only tape one of the magazines; after the first one is out, the other one or two can be easily removed. It is also sometimes helpful to cut out part of the magazine pouch so that the cover on it will get out of the way when you open it up and start removing magazines. This will mean that the pouch needs to be replaced much sooner and you'll need to be sure the magazines don't fall out of the pouch if you forget (or don't have time) to latch it up after removing a magazine.

It is wise to carry empty magazines with you for later use rather than leaving them behind. Don't discard a magazine unless your life depends on the extra speed.

Spare rounds can be carried on stripper clips (ten rounds to a clip) and loaded into empty magazines with a clip guide. Unlike the magazines, these can be carried in cloth bandoliers. Clips are not quick enough to be considered a substitute for magazines, and care must always be taken not to lose the loading guide, or the rounds on clips are little better than if they were in boxes.

Magazines should be carried in pouches with the lips down and, if possible, the ammunition pointing in a direction that will enable you to get it into the rifle without any fumbling. Generally, this means the bullets point out, and the primer side of the rounds is toward your body. A good pouch should be rigid enough to give protection to your magazine's lips.

If a magazine becomes damaged or fails to feed properly, destroy it. Parts should be salvaged from it only if the lips are bent, or if it is obvious what is wrong with the magazine. If you're not sure, discard all the parts. It's better to have only working spare parts rather than a lot of parts that may be defective.

Current magazines used by the U.S. military are the Adventure Line magazines (a name that undoubtedly has caused a lot of snickers on the battlefield and during training exercises) and Colt magazines. Currently, the United States uses 30-round magazines. Colt magazines are good, but they are often plagued with too loose of a floor plate, which can leave the user with an empty magazine in his rifle when the plate slides out and the spring, ammunition, and magazine follower pop out.

All floor plates of magazines (regardless of the brand) should be checked, and they should be taped. The plate may be pulled off part way to be sure the spring is on the outside of the indentation in the floor plate. Tightening the tabs which hold the plate in place may also help. (Be careful, however, not to bend the lips of the magazine when you pound on the tabs.) Also, be positive that you don't tighten up the floor plate so much that it won't slide out without excessive pressure. The idea is to keep it in place until you want to remove it to clean the magazine.

Adventure Line magazines are less prone to plate slipping and are quite good. If you have to pay a premium price for the genuine Colt magazines, get Adventure Line magazines if you're into using your rifle rather than collecting magazines.

There are some magazines that have become collector's items and should never be used if at all possible. They probably won't be encountered, but collector's items occasionally do make it onto the surplus market.

One collector's magazine is the waffle-patterned 20-round magazine originally made for the first military AR-15. Another is the 20-round plastic magazine, not to be confused with modern ones. Both are becoming rare and could be traded for a number of good, modern magazines. If you come upon either one, do not use them. (They were not noted for their reliability.) Among the collector's magazines is the red 20-round blank magazine, which has a spacer bar in its front so that only blanks can be used.

Regular 20-round magazines are available for the AR-15. These are suitable but are nearly as expensive and as large as the 30-round magazines. The added firepower of the 30-round magazine makes it a better buy for most purposes. Your best bet is to stick with the 30-round magazine if at all possible.

The AR-15 Sporter rifles currently are sold with two 20-round magazines, which have a block in them to allow only five rounds to be placed in the rifle (a hunting restriction in many states). If you won't be using the rifle for game hunting or if such hunting isn't limited to five rounds, slide out the floor plate and remove the metal "U." You'll then have two good 20-round magazines.

At the other end of the scale from the 5-to-20-round magazine is the 40-round magazine. These are not being made currently by Colt or Adventure Line, but they are usually nearly as good. (Sterling is one of the companies making fairly good 40-round magazines.) These magazines command a premium price for their extra 10 rounds. They are apt to "high center" if used with a bipod or may make it impossible to engage targets at a high angle when shooting from the prone position. At the same time, a couple of 40-round magazines in the special two-magazine holder arrangements (described below) offer a lot of firepower and might even make it possible to operate without the need of web gear. Sometimes being able to carry everything you need in your pockets and on your rifle is a real plus; for law enforcement officers or those who find themselves in positions where they may have to grab the rifle in a hurry and take off, the 40-round magazine makes sense. They're available from Parellex (1285 Mark St., Bensenville, IL 60106, 312/776-1150) for $29.95 each.

Several companies are also manufacturing plastic magazines for the AR-15. These function very smoothly since the plastic-on-plastic moving parts tend to be self-lubricating. The feed lips on plastic magazines also hold their shape very well and aren't sensitive to bumps or bending of the lips or body of the magazine. Often a serious crunch that would put a metal magazine out of commission will not even hurt a plastic magazine. Since this is how most magazines meet their end, the plastic magazines last quite awhile longer than the metal magazines under most circumstances. Although some users claim that plastic magazines sometimes have trouble with particles of grit getting embedded in the plastic, thereby creating a magazine that works poorly, I've never encountered this difficulty. If you are in an area with a lot of wind and sand, the plastic magazines might be a second choice to the aluminum magazines; otherwise, they seem pretty hard to beat, even though they're more expensive than aluminum magazines at present.

I feel the best plastic magazines are the solid black ones currently available from Numrich Arms (West Hurley, NY 12491, 914/679-2417) for $9.95 each. Clear plastic ones are also available from Defense Moulding (Box 4328, Carson, CA 90745, 213/537-6217) for $9.50 each. Though these magazines are often seen as a plus since you can see how much ammunition is left in them, in fact it is almost impossible to see how much is left in your rifle, since the magazine well of the AR-15 covers up about fifteen rounds worth of ammunition. If you have to take the magazine out to check it, you can usually have a good idea of how full it is by its weight. The clear magazines also have a little problem with reflecting light.

Drum magazines for the AR-15 are found on the market. In general, the drum or snail magazine has never been too satisfactory throughout the twentieth century on any weapon. They work but are as sensitive as regular magazines to bumps, and they are often much more expensive to replace and build. The same seems to be true with the AR-15 drum magazines on the market: they are expensive and just as sensitive to mistreatment as regular magazines. A dent on an aluminum 30-round magazine means you're out of several bucks; a dent on a drum magazine means you're out of a bundle.

A better alternative to increasing the ammunition mounted in the rifle is the practice of putting several magazines together. Do not place magazines together in such a way that one is pointing up while the other points down. This ruins the lips of the one pointing down, and may get enough dirt into the magazine to jam your rifle when the bad magazine is shoved into it.

The way to carry several magazines joined together is with them all pointing in the same direction. Use a spacer between them if you tape them together.

A better alternative to tape and a spacer is a spring clip similar to that used by the Israeli Army for their AR-15. These hold two magazines together. One is placed in the rifle and, when it is empty, it is released, with the full one inserted in its place. This is very quick, and prevents fumbling with pouches or even carrying them if a limited need for the weapon is foreseen.

Spring clips often are not as tight as they should be, however, and a loose one may cause you to lose a magazine somewhere along the line if you're moving about quickly. Usually spring clips should be augmented with a little tape.

The best method currently available for holding magazines together are connectors which hold two magazines parallel to each other. These are available as the "Mag-Pac" from A.R.M.S. (230 W. Center St., W. Bridgewater, MA 02379) for $12.75 each, and the "Magazine Connector" from Choate Machine and Tool (Box 218, Bald Knob, AR 72010) for $15 each.

Each of these holds two magazines very rigidly at just the right space apart to work in the AR-15. If you want to carry two magazines at a time in your rifle, this is the ideal way to do it. When using these, be sure to start with the spare magazine on the selector side of the rifle; otherwise, it is possible to block the ejection port cover so it cannot open. When you place the empty magazine on the ejection port side of the firearm, be sure the port cover is down.

Since double magazines are hard to carry and store unless they are in the rifle, you may find that you can easily use only one group of magazines which are joined together; maximum capacity can be gained with 40-round magazines joined together or through the use of three 30-round magazines joined together (even 40-round magazines can be used if you don't mind the excessive weight). It is hard, however, to conceive of a time this

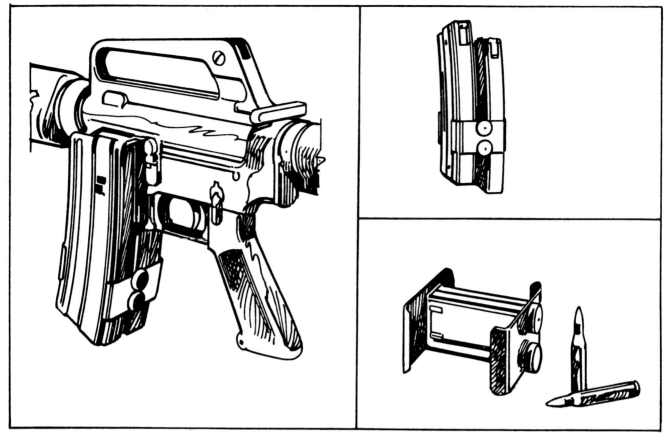

The ARMS double-magazine clamp is shown at left; the Mag-Pac is illustrated at the right.

would be practical unless you are using an AR-15 in the full auto mode as a SAW.

Another interesting solution to carrying several magazines at once in the rifle is the Redi-Mag. This allows you to carry regular magazines with a spare one to the selector side of the rifle. It has one plus: Regular magazines can be placed in it easily so that regular web gear can be used to carry the spares.

The Redi-Mag weighs eight ounces and is attached to the rifle with two clamp screws. The rifle is not altered; a clever latch allows the magazine release on the AR-15 to also release the spare magazine when the empty is released. This makes it very quick and simple to change magazines.

When the full magazine has been placed into the rifle, the shooter can slip a new spare magazine into the holder by just pushing it into place. The Redi-Mag is available from J.F.S., Inc. (P.O. Box 12204, Salem, OR 97309) for $39.50.

Double (and especially triple) magazine configurations are *not* recommended for general use. They put a lot of extra weight on the rifle, make it awkward to carry for extended periods, and make it harder to keep on target. Be sure you really need

them before you go that route.

Magazines should occasionally be taken apart and cleaned. Be sure you use a light coat of lubrication inside the magazine rather than a lot of lubrication—oil can deactivate ammunition and attract dirt. The best lubricants for magazines are those which don't contain any oil at all but are instead based on graphite or similar materials.

New magazines should always be carefully tested to be sure they work. Magazines and rifles don't always work together, even if they are both in good working order. Just because the magazine works or is new does not guarantee that it will work when you're shooting with it. It is better to find that out before you have a lot riding on its working.

New metal magazines often come from the factory with a sharp edge on the upper rim of the magazine over which the cartridges are stripped and pushed into the chamber. Often cartridges will hang up on this sharp edge. Though this will gradually wear off, it is often a good idea to gently file off the sharp edge. Remove only enough metal to get rid of the burr, making it slightly smooth. (Another practice is to be sure that any reloaded rounds you use are chamfered—the edge of the brass is rounded off—so that they won't hang up on the feed edge of the magazine.)

Fully loaded magazines can be stored for years without the spring "setting," just as the hammer and trigger spring in the rifle will not set even if they are left in the same positions all the time. However, they will work a little better if once in a while the springs are allowed to move to their fully expanded point. Do not worry about leaving the magazines loaded for six months to a year, since they will still work fine if they worked well before.

The magazine is one of the most important parts of your rifle; treat it with the care a very important part deserves so that it won't fail you when you really need it to function reliably.

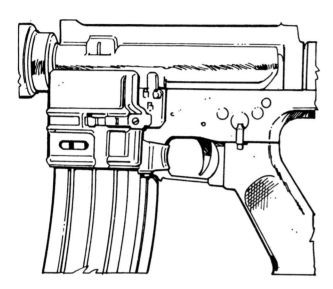
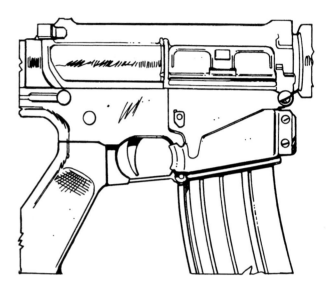

Redi-Mag system for holding a spare magazine

3. Ballistics and Ammunition

You need to forget a lot of the battlefield tales (second or third account) you've heard about the AR-15 rifle and its .223/5.56mm ammunition. Truth or fantasy, such stories may just be exceptions to the norm—if indeed they ever happened.

There are a lot of documented facts and studies on ballistics which deal both with the bullet in flight as well as what happens when it finally "gets there." Facts are something to stake your life on, and your life may be on the line someday when your AR-15 is all that stands between you and death.

First of all, virtually any bullet can be lethal. There is even a documented case of a child who was killed by an accidental chest wound from a 5mm (.20 caliber) air rifle! (See the *Journal of Forensic Sciences,* Vol. 21, No. 3, July 1976, if you have any doubts.) Any round can be dangerous if it is traveling faster than 350 feet per second; at that speed it can penetrate skin and bones. A projectile hitting the eye can travel even slower and still cause severe injury. Generally, military planners consider the minimum amount of energy required to put a man out of action (not necessarily to kill him) to be 108 foot pounds of energy. The 55-grain 5.56mm has that much energy well beyond 500 yards.

Likewise, almost any round can fail to quickly take an attacker out of the fight. All rounds can kill, but they may not kill or incapacitate quickly enough. For example, there are documented cases of criminals being hit in the chest by full loads of 12-gauge shot who still stay on their feet and flee! Any round can be ineffective. Keeping that in mind, you should be ready to rule out the few stories that might tempt you to draw false pictures of how effective, or ineffective, any round may be.

No two bullet wounds are alike. In real life, there are just too many variables. Bullets and ammunition will vary; the angle and range of the shot is different; and the health, age, drug use, and mental/hormonal state of the shooting victim will all work together to decide the round's effectiveness. Some people will be bowled over when hit by a .22 LR bullet, while others will be unaffected for several minutes after being hit by a large-caliber or high-velocity bullet.

Though studying records of actual shootings may shed some light on bullet effectiveness, the conclusions of limited studies are not always cut-and-dry. Studying actual cases only begins to make real sense when the numbers examined approach the hundreds of thousands. Few people have access to enough actual shooting information (especially from rifle caliber wounds) to draw realistic conclusions about how effective any round can be.

With these facts in mind, let's look at the actual history and studies that led up to the development of the 5.56mm round.

HIGH-VELOCITY BULLET USE BEFORE AND DURING WORLD WAR II

Throughout the history of small arms, the trend has been toward ever smaller and faster projectiles. This is partly due to the logistics of carrying equipment and supplying troops; the smaller the infantryman's ammunition, the cheaper and easier it is to keep him supplied.

But there is a point of no return: The round must be capable of taking an enemy soldier out of the battle rather than just inspiring him to fight with anger. Thus, militaries have had to split the difference between size and lethality of rounds.

During the turn of the century, smokeless powders came into use by the major military powers, and World War I saw the introduction of high-velocity, jacketed bullets. These types of bullets were smaller and much faster than the bullets used in previous wars.

Doctors treating the victims of the battlefield soon noted certain bullet characteristics: The pointed bullet had a center of gravity toward its rear portion. This caused the projectiles to be easily upset upon impact so that they often traveled on an axis transverse to the flight of the bullet, especially with an oblique impact with human flesh. The wounds, however, were much different from those caused by earlier firearms in other ways as well.

Early studies done during and after the First World War showed that the bullets from the smokeless weapons were extremely destructive to all living tissues at close range. (The bullets tumbled at long ranges, creating extensive puncture wounds.) The bullets took on an explosive effect at close ranges, causing wounds the like of which the war doctors had never seen. Injuries were no longer limited to the tissue immediately surrounding the bullet track since damage radiated for some distance. Though the entrance wounds caused by the bullets appeared small, the doctors found the following effects:

- Limbs were shattered and muscles split from each other along their intermuscular planes.
- Vessels burst far away from the location of the wound.
- The brain could be pulped by a bullet hitting facial bones.
- The spinal cord could be disintegrated while the vertebrae showed little outward damage!
- Tissues which were in a bony or tough container often actually ruptured in an explosive manner.

Writers and those who demonstrate the effects of high-velocity bullets often shoot at jugs of water or the like, which behave much like some of the body's organs. Since liquids cannot be compressed, the impact of the bullet causes the force to go in all directions in the liquid, thereby exploding the container of water. These demonstrators point out that the body is largely water, and a human body, therefore, also explodes with the impact of such rounds.

In fact, the human body is not like a jug of water (or we would slosh when we moved), but each individual cell in the path of the bullet does react like a miniature water container when it is struck by a high-velocity round. Organs like the lungs, which are full of gas spaces, show little damage from the round. Largely confined, watery organs (like the brain, which is encased in fluid covered with a bone container) will often demonstrate many of the explosive effects of the water container—though not to such a spectacular extent. The damage caused by the fluid pressure to the tissue around the wound is commonly referred to as "hydrostatic shock."

As early as 1898, scientists understood much of the mechanism that created the terrible wounds seen in World War I. The kinetic energy of any missile is imparted to tissue when it strikes, and the tissue in turn bounces about—in effect, "ringing" from the blow. The movement of the tissue and bullet creates a vacuum into which air rushes. This in turn creates a temporary cavity thirty to forty times the size of the projectile, and extreme pressures (up to 1,500 pounds per square inch) are present for a few milliseconds.

World War I moved these theories of high-speed wounds into the everyday life of many soldiers. Government studies of the bullet wounds concluded that the new bullets created a high concentration of air which followed behind the bullet and helped make the wound much worse than it would have been without the effect (just as the scientific theories of the day had suggested). Though the temporary cavity existed for only a fraction of a second and then the wound shrank back to its final size, the damage was done to much of the tissue. Many cells some distance from the final wound channel were killed. Though a wound appeared small initially, it would be quite large in a matter of days as the body rejected dead tissue.

The rifles used in World War I had an average velocity of 2,300 feet per second. Modern studies show that 2,000 feet per second is the point at which a bullet takes on lethal effects, which are created by the gas following the bullet into a wound.

Though World War II was fought with basically the same small arms' rounds as were used in World War I, many countries by the end of the war were looking for ammunition tailored to the realities of combat.

The British Ministry of Supply created the "Small Arms Ideal Caliber Panel" which concluded that the .303 was just too much of a good thing and that a much smaller round might be used with a 600-yard maximum range. The panel felt that a short round with a .270 caliber would be ideal for combat situations. (Unfortunately the U.S. military was still in love with the .30 caliber and browbeat the British into abandoning their search for a better, smaller round.)

Probably much of the "love" for larger caliber weapons dates back to the experimental studies done before the U.S. Army went with the .45 caliber for their pistols during the early 1900s. These tests, in addition to combat experiences, showed that the .45 caliber was about the minimum size for slowly-moving bullets (under 1,200 feet per second) to be quickly effective in combat.

This conclusion was supported by historical trial and error: In the days of smokeless powder, high velocities could not be achieved, and effective firearms therefore had to have very large calibers. The U.S. standard sizes were .69, .58, .50, and .45, with the .30 caliber rifle coming onto the scene only with the invention of smokeless powder.

Later experiments, often done on drugged goats and pigs, showed that the speed of a projectile made a difference if the bullet traveled faster than the speed of sound—approximately 1,130 feet per second (fps), though the speed varies according to temperature and atmospheric conditions.

When the speed is increased over 2,000 fps, the round takes on the lethality evident in World War I. Bullets traveling over this speed are classified as high-velocity projectiles. (The reason for failure of the .30 Carbine round when used in combat is that the round is too small in diameter—under the .45 minimum—and travels at 2,000 fps only near the muzzle, quickly losing speed within normal combat ranges.)

Strangely enough, prior to World War I, the militaries of the world had agreed to limit themselves to solid-pointed bullets in combat, avoiding poison, soft-point, or hollow-point bullets in warfare since such rounds were considered inhumane.

At any rate, the practice is still to limit bullets to those having a solid metal jacket. This is somewhat ironic since little was known about the mechanism of high-velocity bullets when the agreements were made. Consequently, the pointed bullets which were allowed to be used have caused the types of wounds which the signers of the agreements probably thought they were preventing.

The U.S. Law of Armed Conflict still contains the following: "Don't alter your weapons or ammunition to increase enemy suffering. . . . ammunition issued to you (is) legal according to international law. . . . Such 'tricks of the trade' as using 'dum-dum' rounds (bullets with altered tips) or using hollow nose bullets are strictly forbidden."

Thus, while soft-point or hollow-point bullets were banned, a "loop-hole" in the accords allows the use of an unstable bullet, which may possibly be more lethal than the slow-moving, soft-pointed bullets the document signers considered too inhumane to use.

THE BIRTH OF THE 5.56mm BULLET

By the late Fifties, the outcome of the ALCLAD (which took into consideration millions of medical records of combat wounds), SALVO, SPIW, and SAWS studies all pointed toward the need of a round smaller than the .30 caliber 7.62mm NATO, which the U.S. military seemed to want to use. The U.S. military, however, started asking for the development of such a small, more efficient round.

The designers employed by the Armalite Corporation decided to chamber their new AR-15 rifle for the small bullet the U.S. military wanted developed for possible use in combat. The proposed small cartridge with a high-speed bullet would be capable of creating devastating wounds at combat ranges, while being cheaper to produce and easier to move logistically.

The cartridge originally used for the AR-15 was created with .222 Remington brass and a larger .224 diameter bullet (the common size for most of the bullets in the .22 centerfire class).

Stoner, who headed the AR-15 project, was aware that such a round would be only marginal if it were stabilized too much. If too stabilized, the bullet might not always tumble when an adversary was shot. Under such conditions the bullet would still create a nasty wound because of its high velocity, but it would pass through the enemy, rather than tumbling and "dumping" its momentum. Too stable of a bullet would have greatly reduced lethality.

To ensure the bullet would tumble under most conditions, a very slow 1 turn per 14 inches of barrel was used for the rifling in the AR-15 barrel. Such a slow twist creates a bullet with a large yaw. from the spinning axis at the center of its path toward its target. When a bullet has a large yaw, it is somewhat like a toy top with a wobble: As the bullet moves down range, its point makes a small spiral through the air around the center of the bullet's mass. Since the point of a bullet with a large yaw does not hit its target head on, it will have a tendency to tumble on impact as its tail continues forward while its tip is deflected to the side by its spin. As the rate of spin increases, the yaw decreases and the bullet is more stable. If the rate of spin is small, the bullet becomes unstable and slightly less accurate.)

At this point it should be noted that stories of the early 5.56mm bullets tumbling through the air were totally false, though the bullet had a large yaw that often made its silhouette on a paper target have a keyhole shape rather than the circular shape created by other normal and stable bullets. A bullet would have to be fired from a smooth bore in order for it to tumble head over tail in flight.

In order for the 5.56mm bullet to be above the 2,000 fps threshold throughout combat ranges, the bullet also had to have a very high muzzle velocity: over 3,000 fps. (This was the reason for the switch to a larger case by Stoner's designers: There wasn't enough room in the original case for the powder needed to boost the bullet to the high speed it needed.)

The 1-in-14 twist of the first AR-15s gave the bullet fired from the early versions of the rifle a deadly tendency to upset when it hit a target. In test animals (and later human targets on the battlefield) the bullet usually tumbled end over end after entering the body. Its tumbling, coupled with the large amount of air sucked into the temporary cavity created by the bullet's impact, created terrible internal wounds. Such wounds were much worse than one might expect from the small entry wound. Rather than having wound tracks the diameter of the 5.56mm bullet, the wound's width below the skin was often equal to the length of the bullet.

The twist rate was later changed by military planners to make the bullet more accurate in cold climates. Even so, the bullet still was quite unstable compared to many other rounds. Some lethality was sacrificed but not an excessive amount, especially since much of the bullet's ability to wound is due to its high speed rather than its tumbling ability.

The tumbling bullet and air sucked into the wound were not the only sources of damage.

Secondary fragments of bone or other hard tissues often were imparted with enough kinetic energy by the bullet's momentum to cause the fragments to become secondary missiles. Often these fragments caused as much damage as the bullet. For example, if a pistol bullet strikes a glancing wound to the skull, the victim will receive a serious cutlike wound, but his skull will usually remain intact and his brain protected. The same type of injury from a 5.56mm round may outwardly appear to be the same, but it can, in fact, be fatal. When the 5.56mm bullet hits the skull and glances off, it imparts enough energy to crack out a small key-shaped chunk of bone, which in turn flies into the brain, often killing the victim. An even worse effect can result with bullets having a high yaw (as is the case with the 1-in-14 and 1-in-12 twists); the bullet may hit the skull when the pont of the bullet is pointing downward. In such cases, the bullet rapidly digs into the bone. Along with bone fragments, the bullet is deflected inward, thereby tumbling through the brain.

Bullet wounds in other parts of the body also often have bone-fragment tracks radiating out from the site of the primary wound channel. The wounds may outwardly look small, but extensive internal damage will be present.

Many people do not realize how serious the wound from a 5.56mm bullet can be. If a shooter fires a 55-grain bullet from a 1-in-12 twist AR-15, the temporary wound channel that the bullet and the air sucked into the cavity create would be comparable to that made by a slow-moving pistol bullet thirty to forty times its size. If the bullet hit a large bone, secondary wound channels similar to those produced from a shotgun would branch out away from the wound.

Comparing this wound to that produced by a .45 is also interesting: The 5.56mm bullet would make a temporary wound about four to eight times as large as that produced by a .45 ACP round! And though the large channel only lasts for a few fractions of a second, it is long enough to kill most of the tissue in the area. If the jacketed .45 is effective (and it is generally agreed that it is), then facts would seem to indicate that the 5.56mm would have to be considered a lot more effective as a combat round.

The damage doesn't end there, though. The 5.56mm round will also tumble if fired from a 1-in-12 twist. That means that the bullet would cut a swath through the wound channel which would be much worse than that of a .45. A slow-moving pistol bullet pushes tissue out of its path; the tumbling rifle bullet cuts tissue or pushes it ahead of the bullet. The wound channel of the 5.56mm would thus slash through the victim's tissue, severing nerves and blood vessels and breaking bones.

Following is a portion of a Congressional transcript during which Representative Ichord questioned Eugene Stoner during Special Subcommittee meetings on the M16 (the military version of the AR-15 rifle):

Representative Ichord: One [soldier] ... told me that he had shot a Vietcong near the eye with an M-14 [7.62mm NATO bullet] and the bullet did not make too large a hole on exit. [Later] he shot a Vietcong under similar circumstances in the same place with an M16 and his whole head was reduced to pulp. . .”

Eugene Stone: That is the advantage that a small or light bullet has over a heavy one when it comes to wound ballistics. . . . bullets are stabilized to fly through the air, and not through . . . a body. . . . When they hit something, they immediately go unstable. A .30-caliber bullet might remain stable through a human body. . . . a little bullet senses an instability situation faster and reacts much faster. This is what makes a little bullet pay off so much in wound ballistics.

Interestingly, when the USSR introduced their new .22 caliber in Afghanistan, many reports surfaced about the “super bullet” which they had developed. In fact, rounds recovered from the battlefield proved that the round was inferior to the 5.56mm.

Before anyone knew much about the round, stories surfaced concerning the wounds it caused. Such wounds were not unlike those reportedly made by the AR-15 round when it was first introduced to the battlefield. Reporters speculated that the Russian bullets were poisoned or traveled at 4,000-plus fps. In fact, the damage was caused in the way all high-velocity rounds since World War I produce their deadly effects. The apparent poisoning around the wound was created because the tissue around the bullet's channel had been killed by the bullet/air impact. Though the skin around the wound looked healthy, it in fact was dead and slowly rotting around the wound, thereby making it appear the wound channel was subject to poison. (Interestingly, the Russian bullet has an enclosed hollow point which causes it to be even more unstable than regular pointed, solid bullets. This causes the round to tumble more readily than a solid bullet of the same size.)

Like the USSR, a number of countries soon switched to smaller caliber rounds as the rounds' deadly effects became apparent on the battlefield. NATO countries adopted the 5.56mm round, as did large numbers of small countries, including South Africa and Israel. (Israel produces a .308 version of their combat rifle for export. Though Israel expected to be fighting in desert areas, it still embraced the 5.56mm round after several wars with Arab neighbors.)

In addition to its lethality, the 5.56mm round had a lot of other advantages as well:

- It was only about half the size of the 7.62mm NATO round.
- Battle rifles for the 5.56mm were cheaper to make.
- The rifles for the 5.56mm round could be much lighter than those for the 7.62mm round (which coupled with the weight of the round itself meant that a soldier could carry three times as much ammunition as a soldier with a heavier round and rifle).
- The 5.56mm's lower recoil made it possible to control lightweight weapons during automatic fire (for followers of the SALVO project).
- The round allowed quick recovery from a single shot so that the soldier could re-aim more quickly (for those of the aimed-fire-not-auto school of thought).
- The round speeded up the training of troops so that they could become proficient shooters more quickly.

DEVELOPING AND TESTING OTHER ROUNDS

Though the 5.56mm round first developed for the AR-15 soon became the standard 5.56mm round for many of the rifles in the Free World, efforts were soon made to develop an even better round.

One thrust was to develop a more “humane” bullet, while another push for change came from military leaders who still wanted a round capable of having a minimum usable range of 800 to 1,000 meters. Though endless studies have proven that the average “kill” is made at under 100 meters and that soldiers in combat rarely can see a target far enough away to use long-range fire, several good arguments point out that tactics might change and the round might double for use in light machine guns if it had a better range. A supply problem caused by the need of two different calibers for troops in the field would thus be alleviated.

From 1976 to 1979 NATO started testing out several calibers of rounds, including several types of 5.56mm, for a standard round which could be adopted by their armies. During these tests, some evaluation of weapons was also made, although all countries agreed that a standard rifle would not be adopted by all NATO members.

The main thrust of the tests was to find a common small round for their assault rifles. Acceptance of such a round by all NATO members would greatly aid their goal of using interchangeable ammunition. It also offered a chance to develop a superior 5.56mm round to replace the old U.S. standard round designated as the M193. (The M196 was the tracer counterpart.)

The bullets for the test were the U.S.-developed XM777, a French steel-cased round, and the Belgian SS109 (which Belgium had been working on for use in its own rifles). Two other rounds were submitted but soon removed from the tests: the German 4.7x21mm caseless round (which had cook-off problems) and the British 4.85mm round.

A number of rounds and rifles currently being used by NATO were included in the tests as standards of reference, while others were more experimental. These rifles included the M16A1, the French FAMAS, the Belgian FNC, the German G3 (the automatic version of the H&K 91) and G11 (designed for 4.7x21mm caseless rounds), the Dutch MN1 (the Dutch-manufactured Galil), the Belgian Minimi, and the British EWS (originally in 4.85x49mm and later rechambered for 5.56mm). (Interestingly, it was noted by the NATO testers, who had no vested interests in the U.S. military's AR-15, that the M16A1 was the most reliable weapon in the 5.56mm caliber.)

The SS109 proved to be the best round during the tests and was adopted by NATO in October 1980. (The round was designated the M855 by the U.S. military, and its tracer counterpart became known as the M856.) The round has the same dimensions as the M193, but it has a bullet of 62 grains rather than the 55-grain bullet of the M193. The bullet has a steel insert which sticks out of the tip of the bullet, thereby giving the round abilities that are superior to the larger 7.62mm NATO round. While the 7.62mm bullet only penetrates a U.S. steel helmet out to 700 meters, the SS109 bullet perforated it out to 1,300 meters! The smaller bullet is likewise able to penetrate more armor plate than the 7.62mm NATO bullet.

The US XM777 was superior in its abilities to wound, but it did poorly in penetrating steel plates, helmets, etc., at long ranges. It appeared that the round might also be expensive to produce. The SS109 was thus chosen as the second standard NATO round (the 7.62mm NATO being the first) even though it probably has less lethality at combat ranges and would require the rebarreling or replacement of French, United States, and Netherlands rifles before they could effectively use the new SS109 round.

To date, two important questions remain unanswered: Which is more important, increased armor penetration or a round which tumbles on impact? Do tumbling rounds really produce greater casualties when modern armies are developing better helmets and ballistic armor? Probably only battlefield or extensive animal tests will produce the answers to these questions.

Another question of modern tactics also remains unanswered: Are high rates of fire the most effective way to deal with an enemy? The reduction of caliber through history has been coupled with the increase of rounds expended in combat. During the American Revolutionary War, soldiers fired seventeen shots for each enemy casualty. The Vietnam War saw 50,000 rounds fired for each casualty produced. It would seem that changes in tactics and doctrine must be made which allow more emphasis on the quality of fire and reduction of expended rounds.

While working on this book, I studied a number of combat tapes. From Vietnam, to El Salvador, to Lebanon the same picture was continually evident: Riflemen would fire entire magazines toward enemy positions. Usually the weapons were fired without aiming, and often the shooter did not even stick his head over his barricade. Instead, he held his gun over the barricade and fired in the general direction of the enemy. Surely there is a better way of fighting. Perhaps better training and the three-round burst which will be on new military AR-15s will help.

4. Functioning

The AR-15 operates in a manner very similar to most other automatic rifles. The main difference is in its gas system, which does away with the piston/rod arrangement used in the majority of rifles. Its system funnels the gas from the gas port in the barrel down through a tube and into the bolt key of the bolt carrier in the upper receiver. Here the gas forces the bolt to unlock, extract the empty brass, and travel back toward the rear of the receiver.

This system makes for a much simpler rifle and lightens the weapon. It can also create a lot of problems with dirty or slow-burning powder as unburnt particles are blown directly into the bolt area where they may accumulate if the weapon is not carefully cleaned regularly.

Most versions of the AR-15 function in either a semiauto or automatic mode. Except for a few differences between these two modes of fire, the weapon functions almost identically when fired either way. The AR-15 which fires only in the semiautomatic mode generally lacks the auto sear and has a disconnector modified to prevent it from being disengaged if the selector is in the "auto" position.

The commercial AR-15 is further modified so that if the disconnector is removed the hammer will be caught by the firing pin. (This is why the commercial AR-15 hammer has a notch on its face and the bolt is cut away under the firing pin.)

There are seven distinct events that occur when the rifle is fired (though several of them may overlap): the firing of the round, unlocking of the bolt, extraction of the round, ejection of the round, the cocking of the hammer by the bolt carrier, the feeding of a new round into the chamber, and the locking of the bolt.

In semiauto fire, the events happen in the following manner.

Before the rifle can be fired in the semiauto mode, a round must be placed in the chamber, the hammer cocked, and the selector in the "semi" position (or "fire" with semiauto-only rifles). When the trigger is pulled, it releases the hammer, and the hammer is propelled forward by its heavy spring. The face of the hammer slams into the rear of the firing pin, and the energy from this collision is transferred down the firing pin onto the face of the primer.

The primer ignites, causing the powder in the cartridge to ignite. The gases created propel the bullet down and out of the barrel.

As the bullet passes the gas port in the top of the barrel, gas from the burning powder of the cartridge begins to be bled off into the gas tube.

This high-pressure gas quickly moves down the gas tube toward the action of the rifle so that the gas forces the bolt carrier back shortly after the bullet exits the barrel. After moving backward a fraction of an inch, the bolt carrier forces the cam to twist the bolt, thereby unlocking its lugs from the barrel extension.

Once the bolt is unlocked, it and its carrier are pushed backward by the last of the gas pressure from the gas tube. As the bolt moves backward, its extractor holds the extraction groove of the cartridge. The ejector pushes at the empty cartridge, forcing the mouth of the cartridge to rotate toward the ejection port side of the upper receiver. The bolt carrier quickly travels backward, cocking the trigger. As the cartridge passes the ejection port, the brass flies out.

If the firer has not released the trigger by the time the action cocks the hammer (and it is almost impossible to release it that quickly), the hammer is trapped by the disconnector spur which engages the hook at the rear of the hammer's neck. This will prevent the hammer from later following the

COCKING

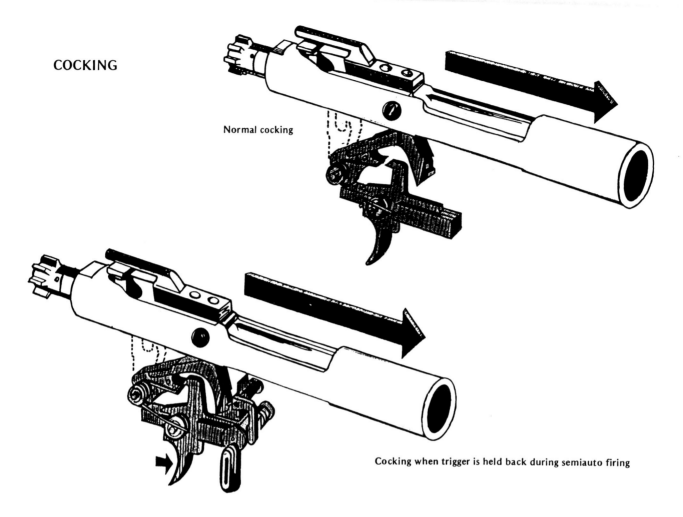

Normal cocking

Cocking when trigger is held back during semiauto firing

bolt carrier forward. When the shooter releases the trigger, the disconnector also releases the hammer so that it is held by the trigger's nose as it was before the shot was fired.

The buffer is full of loose projectiles (discs or shot) which are independent of the outside of the buffer. Thus, when the bolt carrier hits the face of the buffer, the outside of the buffer is propelled backward after soaking up some of the carrier's energy. The loose pieces in the buffer, however, remain stationary momentarily. In so doing, they stop the outside of the buffer as it imparts some of the carrier's momentum to the inside pieces of the buffer. This is what creates the "ta-whap" recoil of the AR-15. It takes a fraction of a second for the carrier, the buffer, and the inner buffer parts to get going in the same direction!

The carrier and buffer start moving back together against the buffer spring. The spring quickly soaks up the energy of the moving parts and then pushes them back toward the front of the rifle.

As the carrier moves back into position, the bolt strips off another round from the magazine. As the

bolt moves forward the last half an inch of its travel, its cam causes the bolt to twist and lock behind the cartridge. The extractor is forced into the cartridge's extraction groove, and the ejector spring is compressed.

Now the rifle is ready to fire another round when the trigger is again pulled.

During automatic fire, the operation is basically the same except that the disconnector is disengaged by the selector (placed in the "auto" position), and the automatic sear comes into play.

In auto mode, the carrier moves the hammer rearward. The auto sear catches the top hook on the rear of the hammer and holds the hammer back (rather than the disconnector holding it back as it does when in the semiauto cycle). When the bolt locks up with the cartridge in the chamber, the rear of the bolt carrier levers the auto sear and the hammer is released. If the shooter continues to hold the trigger back, the hammer falls against the firing pin and fires the round. If the shooter has released the trigger, the base of the hammer is caught by the trigger in the usual manner.

FIRING

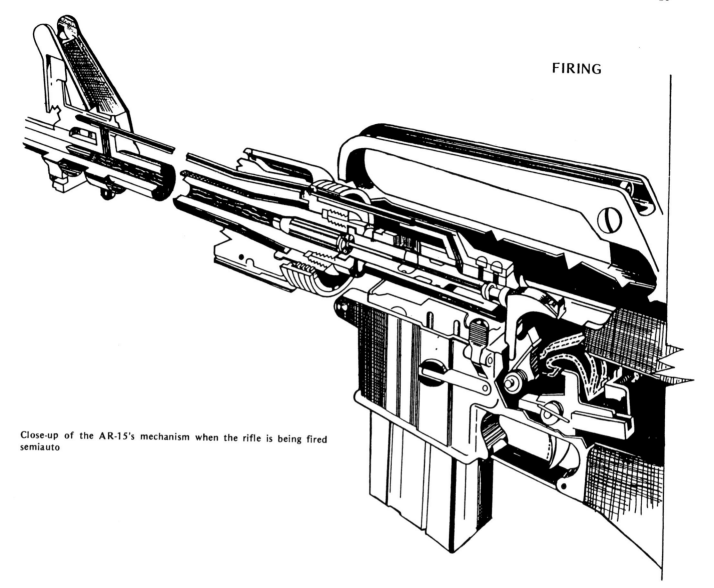

Close-up of the AR-15's mechanism when the rifle is being fired semiauto

As long as the trigger is held back and the magazine is full, the rifle continues to fire in the automatic mode.

When the magazine is empty in either mode of fire, the magazine follower pushes the bolt catch upward. The bolt catch then keeps the bolt and its carrier from traveling forward after cycling back.

5. Learning to Use the AR-15

Once you, or someone you're teaching, have learned how to operate the action, checked the rifle to be sure everything is in order, and become familiar with how your rifle works, you're ready to begin shooting it.

When firing the rifle, it is important to use ear protectors. This keeps you from damaging your ears and—just as important from an accuracy standpoint—will help to keep you from developing a flinch.

Flinching is caused when shooters anticipate the recoil of the rifle. This movement causes the shot to be very inaccurate. Once a shooter gets into the habit of flinching, it will be quite difficult for him *not* to flinch.

It is possible to discover whether you (or some other shooter) have a flinch. To do so, mix a dud round in with your other practice rounds when you load up a magazine. (No fair keeping track of where it is in the magazine!) When it fails to fire, you'll know if you have a flinch. Anyone with a flinch will move their rifle as if it had recoiled from a blast even though the round didn't fire.

If you reload, it is not much of a problem to create a dummy round. Just resize an empty brass, leaving the spent primer in it, and seat a bullet in the brass. Mark the round so it won't be confused with live rounds, and be sure not to use the brass for loading later. It may become too short for the rifle after the round has been slammed by the firing pin against the front of the chamber without being fire sized by the ignition of powder.

One way to mark the round is to use lacquer or fingernail polish around the bullet. Perhaps a better way is to drill a large hole through the side of the case. This makes a more permanent identification, and the brass can't be reloaded by accident.

If you have a .22 adapter for your AR-15, using it when learning to shoot is very useful if you have never fired a rifle. The .22 cuts the recoil and report down to the point where the shooter won't be startled if he has had some training and wears ear protectors.

The best way to place single practice rounds in the rifle is to leave an empty magazine in the rifle with the bolt locked back. If your fingers are not too large, you can reach through the bolt and place a round on the top of the magazine and push the shell down into it. With the round locked in the magazine, the action can be released and the round will be chambered without problems. Do not drop the round into the chamber and release the bolt. The round may be fired since the bolt won't be slowed down enough to keep the floating firing pin from igniting the primer.

Another route you can go if you reload is to use light or "squib" loads. These are rounds with only small powder charges. Though they won't function the action on the rifle, a number of the rounds can be placed in the magazine, and the rounds can be fed through by cycling the action as each round is needed. Some people prefer single shots for beginners since the beginner can't fire a second round accidentally.

Squib rounds can be made with 45-grain Hornet bullets (which are .224 inches in diameter just like the 5.56mm bullets) or with regular 5.56mm/.223 52-, 53-, or 55-grain bullets. These are loaded over light charges of pistol or shotgun powder.

Three different squib loads which can be used for training consist of a 45-grain bullet over 6 grains of Unique (which gives a muzzle velocity of around 1,900 fps), 52- or 53-grain bullet over 6 grains of Unique (giving around 1,900 fps), and a 55-grain bullet over 8 grains of Unique (giving a muzzle velocity of about 2,100 fps). It is also possible to create other loads with cast bullets and load them over 3 to 5 grains of Unique. The Lyman 49-grain lead bullet from their mold number 225415 is good for this. As long as the velocity is kept low, the lead bullet won't foul the bore of the rifle.

Like the dummy rounds, the brass that has been used for squib rounds should not be reloaded to full charges since such use would be dangerous.

When making squib rounds, be sure you do not drop too low with your powder charges since a bullet may get jammed in the barrel. If a bullet does lodge in the barrel, a second one fired down the tube could be a tragedy. If a bullet gets stuck in the bore, lead bullets can be tapped out with a cleaning rod, but jacketed ones must be removed by a screw-type tool—a job best left to a gunsmith.

Dynamit Nobel is producing small plastic training rounds for the U.S. military. These 5.56mm rounds consist of a plastic bullet and case which are molded of one piece of plastic with a preset breaking point which allows the two to separate upon firing. A primer is mounted in the small metal base, and a noncorrosive powder charge is enclosed in the round. The plastic bullet has a small metal insert, which can be used to contain chemicals for the tracer rounds. The plastic bullets have a high initial velocity because of their light weight; this speed quickly drops due to air resistance to the bullet. This makes the round ideal for areas where there is insufficient space to normally allow for rifle fire.

There are currently several .22 rifles which out-wardly resemble the AR-15. Often these are sold as training aids to those learning to use the AR-15. In fact, these rifles have different safeties, magazine releases, and sometimes even different cocking devices. Sold under the names of the Concorde "M-16" rifle or the "AP-74 Military U.S. Army Rifle," these rifles only bear a resemblance to the AR-15 and aren't used by the military. They are not of much use for training. These rifles resemble other .22 rifles more than they do any version of the AR-15. For about the same price as these rifles, you can purchase a .22 conversion kit for an AR-15; the kit is a much better training aid and is as reliable in functioning as the .22 rifles.

If you practice shooting by "dry-firing" your rifle (i.e., cocking it and pulling the trigger so the hammer/firing pin hits an empty chamber), you may eventually damage the firing pin. If you buy a couple of spare firing pins and replace the pin before you use the rifle for shooting, this is no problem. Another solution is to cut out some plastic "primers" from the caps that come on plastic gallon milk jugs, and glue the plastic primers into the primer pocket of an empty brass case. If you grind off the extraction groove of the brass, it won't come out each time you cock the rifle. (If you don't grind off the groove, be sure to mark the brass so it won't be used for reloading.)

Careful practice and training can enable you to become familiar with your rifle so that it functions as if it were a part of you. There is no such thing as too much practice if you may ever need your rifle in combat.

6. Cleaning and Lubrication

It is important to keep any firearm clean so that it will operate at its best. At the same time, many weapons have received undue wear because their owners cleaned them *too* often.

New lubricants like Break Free CLP (my favorite), Tri-Flon, TSI 300, or Mil-L-46000A have changed the chore of rifle cleaning. Break Free can be used to clean and lubricate the rifle in one step rather than using a separate liquid to clean the gun and another to oil it. There are other advantages as well since the new lubricants/cleaners stay fluid over a wider temperature range, and therefore you do not have to use different lubricants for hot and cold weather. Use the new lubricants if at all possible, avoiding the older gun oils.

Excess lubricant can be as bad as no lubricant at all. It attracts dirt, and the grit may cause more friction and wear to parts than they'd get without any oil. After lubricating a rifle, it is a good idea to wipe its outside down with a rag to remove any lubricant that might attract sand or dirt.

Good cleaning kits are available from most military surplus stores. These are excellent since they were specifically designed for use with the AR-15. You should also add a number of cleaning patches, pipe stem cleaners, and the plastic "tooth brush" cleaners if your kit doesn't have them. If your AR-15 has a trap door stock, all of the cleaning gear—along with a small bottle of Break Free CLP—will fit into it. (You should wrap the whole kit in a rag and a rubber band so the gear doesn't rattle.)

Many people shy away from the steel cleaning rods of military cleaning kits because they fear the steel rods may cause extra wear to the bore. In fact, the steel often causes less wear in combat conditions since the soft aluminum or brass rods can pick up grit which becomes imbedded in the soft metal. These particles are harder than the lining of the barrel and can quickly damage a barrel. The steel rods will work well if you do not scrape them against the inside of the barrel. Be especially careful at the muzzle of the rifle so that the rod does not rub against it to ruin its accuracy.

The wire bore brush should be pushed through the barrel from the receiver end of the rifle; be sure to push the brush on through the flash suppressor before reversing directions. If you try to reverse directions in the barrel, you may jam the brush in the bore.

When cleaning the rifle, be sure to avoid using wire brushes on the aluminum receiver; the metal brush will take the finish right off the receiver! Wire brushes should not be used on the Promethium (glow-in-the-dark) sight since it can also be damaged by a metal brush.

When you've finished cleaning and lubricating the weapon, inspect it to be sure that patch or brush bristles are not trapped anywhere in the rifle. Lightly lubricate the barrel and small parts like the windage drum, ejection port cover, detents, charging handle catch, the bolt, or the bolt carrier.

The inside of magazines should be lubricated from time to time, and the outside of steel magazines should be wiped to prevent rust. (Most magazines are composed of either aluminum or plastic, whose outsides only need to be cleaned of dirt.) The new plastic magazines are sealed much better than are the metal magazines; the plastic magazines therefore need less cleaning than metal ones because less dirt gets into the plastic magazines.

If time is at a premium, it is possible to clean and lubricate only the most critical parts of the rifle to keep it functioning well. Since many AR-

15s have a chrome-lined barrel, it is possible to "get away" with limited cleaning without worrying about rust forming inside the bore.

The "quick clean" is done by clearing the rifle; opening up the receiver halves; removing the bolt carrier; and taking out the bolt, firing pin, retaining pin, and cam. Scrape off the carbon and old lubricant from the firing pin and the bolt; clean out the firing pin hole. Clean out the bolt carrier and the inside of the bolt carrier key with a pipe cleaner. Clean the chamber with wire brushes and then a cleaning pad (on the cleaning rod). If you have used cleaning fluid rather than Break Free or other cleaner/lubricants, lightly oil all the parts and the barrel extension lugs. Reassemble the rifle, and you're ready to use it again.

Time permitting, a more thorough cleaning and lubrication of the rifle should be done.

In environments exposed to salt spray, dampness, or high humidity, the rifle should be cleaned at least twice a month even if it hasn't been fired. To prevent rust on steel parts, rifles in drier climates should be cleaned at least several times a year even if the firearm has not been fired.

When cleaning and lubricating the weapon, inspect all the parts to be sure they are not damaged or excessively worn. Sometimes it just takes a little extra work to get a part clean. The rifle's parts must be clean; oiling dirt won't help much.

The barrel should be scoured with a brass or copper bore brush and then swabbed down until the patches come through it clean. Care should be taken to clean and lubricate the locking lugs and the inside and outside of the gas tube area that extends into the upper receiver. Clean out the chamber with the chamber brush; use at least four or five plunging strokes and three full turns inside the chamber. The bore and chamber should look shiny and clean when you look through them toward a light.

The bolt carrier should be cleaned and lubricated, with special care being taken to clean the inside of the carrier key and the inside of the carrier where the bolt rides. The charging handle should be carefully cleaned and lubricated.

The bolt should be disassembled and cleaned out. Be sure to thoroughly clean the rear area of the bolt between the bolt body and gas rings. The inside of the locking lugs and the hole through which the firing pin runs are often full of carbon deposits which must be removed.

The lower receiver, buffer, and buffer spring should all be carefully cleaned of dirt, and the front sight should be also brushed clean.

After all parts of the rifle have been lightly lubricated, add extra lubricant to the bolt, gas rings (which should not be lined up), cam pin, firing pin hole, and push pins. One drop of oil should be put down the carrier key hole, and the lower action lightly lubricated.

Careful cleaning and lubrication of the AR-15 will allow you to have one of the most effective and reliable small arms in the world.

7. Operation

When loading or cycling the AR-15 by hand, remember that malfunctions do occur and the rifle might fire unexpectedly. Keep the barrel pointed in a safe direction at all times. People get killed every day by "empty" rifles or while "checking out" a rifle. Before you load your weapon or dry fire it, decide whether you can live with a bullet hole in the area the rifle is pointed at! This should be remembered in combat, too; the number of troops killed by accidental "friendly" fire is unforgivable.

The floating firing pin of the AR-15 is normally quite safe. But, if it strikes a high or sensitive primer, it can fire the round as it is chambered. A sensitive primer that causes a slam fire won't damage the rifle (since it is designed so that the pin can't reach the primer unless the bolt is locked in place), but the bullet coming out the barrel can cause a whole lot of unexpected damage.

A primer that isn't seated to the proper depth can damage the rifle—and you and/or a bystander—since the primer may be set off by the edge of the bolt as well as the firing pin. The rifle probably wouldn't be damaged beyond repair, but why chance it? If you reload, inspect every round to be sure primers are seated to the proper depth. It's been my experience that you should *never* use rounds reloaded by a friend. If a round malfunctions, you can lose a friendship as well as a rifle, a right eye, etc.

Another thing *not* to do is to pull the charging handle back, lock the action open, drop a round into the chamber, then release the action so that it crashes shut. Chances are good that the round will be fired by the firing pin. During normal operation, the rifle strips cartridges off the magazine. This slows down the bolt enough so that the firing

pin won't fire the round. If you don't have anything slowing down the bolt and bolt carrier, instead of the normal small indent on the primer, you may have a fired cartridge with the brass out to your right and a bullet hole somewhere ahead of you. Be careful!

A final caution has to do with a jammed action. This can be created with oversized ammunition, dirty ammunition, or a fouled chamber. Normally you can prevent a jammed action by not using the bolt assist under any condition. If the round doesn't chamber when the action strips it off the magazine, you're just asking for a jammed weapon if you give a tap to the bolt assist to help the round get into too tight a chamber.

The only time you might get away with using the bolt assist is if you're using the rifle in extremely cold conditions, or if you don't release the charging handle but hold onto it, easing it forward to chamber a round (this allows you to chamber a round very quietly). In such cases a very light push on the bolt assist might be of use.

LOADING THE AR-15

The first step in loading the rifle is to get a magazine full of ammunition. Individual rounds can be chambered by hand in an emergency, but it's a little risky unless you follow the charging handle forward rather than releasing it so that you don't have a slam fire in the manner outlined above.

Magazines can be loaded a round at a time or from stripper clips of rounds. To load the magazine a round at a time, first remove the magazine from the rifle (if it's there) and hold the magazine in one hand. With the other hand, place a round on top of the follower with the bullet end of the round toward the front of the magazine. When it's on the magazine follower (the part that moves down into the magazine), push it down below the lips of the magazine and release the round. It should stay in place on the magazine (if it doesn't, push it farther down into the magazine). Repeat these steps until the 20, 30, or 40 rounds are loaded into the magazine.

Stripper clips can enable you to load a magazine in a hurry if you've got the clips loaded up beforehand, and if you have the clip guide handy. Stripper clips are not a good substitute for charged magazines but can be used to speed things up if the magazines aren't available.

The clips are usually available on the surplus market. (One good source for stripper clips and guides is Sierra Supply, Box 1390, Durango, CO 81301, 303/259-1822. Guides currently cost 50 cents; clips, 10 cents each; and a bandolier with cardboard sleeves, $2.50.) You need a stripper clip guide to use the clips when charging a magazine.

Bandoliers are handy to carry the clips in. Though the cardboard sleeves for the bandoliers only hold two clips, a third can be placed to one side of the sleeve. The guide can then be carried in one of the pockets or—better yet—tied to one end of a nylon cord, and other end of the cord tied to the bandolier.

The clips are filled with ten rounds of ammunition, and the brass tag at each end of the clip is bent to a right angle of the clip itself so that the rounds are locked onto the clip. The clip and rounds can then be carried until they're needed.

When it's time to load an empty magazine, the stripper guide is placed over the top, rear end of the magazine, and a stripper clip is placed in it so that the bullets of the cartridges are pointing in the proper direction.

The rounds are then pushed down so that they bend the brass tag and move off the clip into the magazine. Sometimes this can be done with one smooth motion; often you'll fill the magazine with a series of pushes. Some people find it easier to shove just four or five rounds into the magazine at a time rather than all ten; the base of the magazine can be placed against the chest so that both hands can be used to push rounds off the clip into the magazine.

After several clips have been used to put the proper number of rounds into the magazine, the clip guide is removed (be sure not to lose it), and the magazine is ready to be used.

With the magazine still out of the rifle and the barrel pointed in a safe direction, pull back on the charging handle until it stops and then let go of it. This will cock the hammer so that the selector can be placed in the "safe" position if it isn't already in that position. (In combat, you may wish to omit placing the selector into the safe position.)

Shove the magazine into the magazine well, until the magazine release clicks shut on it. Military trainers generally have you slap the base of the magazine so that you can be sure it's fully seated. In combat, this is a good idea since it is often impossible to hear the clip of the release popping into the magazine, and dirt may make a magazine seem to be in place when it is not. It's good to get into the habit of giving the magazine a rap so that it doesn't fall out of the rifle when you really need it.

With the magazine in place and the barrel pointed in a safe direction, pull back on the charging handle until it stops and then let go of it. This will cause the bolt to strip a round from the magazine and chamber it.

The weapon will now fire if you move the selector from the safe position to a fire position. The selector positions vary from one type of AR-15 to another. On the semiauto Sporter, the positions are safe and fire and the third, auto, position is not available. With the M16 and M16A1, the selector positions are safe, semi, and auto. With the M16A2, the positions are safe, semi, and burst.

If you won't be firing for a while, leave the selector in the safe position and close the dust cover over the ejection port. The dust cover will pop open automatically when the rifle is fired or cycled by hand. The dust cover must be closed manually, however.

When all the rounds have been fired, the magazine follower will engage the bolt catch and hold the bolt and bolt carrier to the rear. This also keeps the hammer back so that it won't fall on an empty chamber.

The bolt catch allows you to visually inspect the rifle to be sure it is empty by looking through the ejection port and seeing the empty chamber and magazine. The bolt catch makes it possible to quickly bring the rifle back into action.

For quickly reloading the rifle, push the magazine release and remove the empty magazine (it may or may not drop out depending on your rifle and the magazine). Push a full magazine into place in the well and slap the magazine's base to be sure its seated properly. Press the bolt release, and the bolt and its carrier will crash forward, chambering a round. You're all set to fire again. If you may not be firing right away, again engage the safety.

There are a number of safety precautions that should be observed with the AR-15. Though many apply to all types of rifles, the AR-15's floating firing pin can cause a number of extra complications. Though the problems are not normally encountered in "real life," the unexpected is possible. It only takes one abnormal, unexpected happening to cause a tragedy with a rifle.

Because the firing pin normally rests against the cartridge when a round is chambered, a blow to the muzzle or dropping the firearm on its muzzle might create enough inertia to fire the cartridge.

Before firing a rifle that has been outside for some time or which may have an obstruction (mud, snow, rain, leaves, etc.) in the barrel, always check to be sure the barrel is cleared. Do this by removing the magazine and cycling the rifle so that the chamber is cleared. Inspect the chamber to be sure it is clear. Lock the bolt carrier to the rear and look down the barrel. If anything appears to be in the barrel, clean it out with a bore brush. A muzzle cap is very useful in keeping "junk" out of the barrel. If you have to fire the rifle in a hurry, you can fire it with the cap in place (provided everything else is as it should be).

The bore of a 5.56mm rifle is so small that it is possible for it to hold a column of water inside it. If you've dropped your rifle into water or have been in heavy rain, it is always wise to remove the muzzle cap (if any), point the muzzle down, pull the charging handle back two or three inches, and allow the water to drain out of the bore for a few seconds. When releasing the charging handle after this, you should use the forward assist if your rifle has one so that you can be sure the bolt has closed. If your rifle does not have a forward assist, visually inspect the carrier and push it forward by putting your finger on the lubrication cutout and pushing toward the muzzle.

Keep your finger out of the finger guard until you're ready to fire. If you have your finger in the guard and fall or get excited, it is very easy to fire the rifle accidentally.

When you fire your rifle, be sure you know what is behind and beyond the target. Though the useful range of the AR-15 is only 500 to 800 yards, its bullets can travel up to three miles and penetrate a lot of material. Additionally, ricochets from a rock or water can cause the bullet to travel in a path different from that at which it was fired. Think before you fire.

Always switch from the "fire" position back to the "safe" if you may not be firing for a while. The rifle is much less apt to be accidentally fired if it's left on safe. You should also develop the habit of moving the safety off or checking it with your thumb every time you get ready to fire.

Use only the best ammunition. If possible, use factory-manufactured ammunition. If you reload, use full-case resized cartridges and be sure the primers are fully seated, bullets tightly seated, and the case free from oil. Never fire corroded, dented, or hot (plus 135 degrees Fahrenheit) ammunition.

If you have been firing extensively and have to stop suddenly, it is wise to eject the round in the chamber within ten seconds and lock the bolt open with the chamber empty. This will prevent a "cook-off" which can occur when the heat of the chamber causes the powder in a cartridge to ignite. If it is impossible to eject the round, there is a possibility of a cook-off for up to fifteen minutes. Keep the rifle pointed as if it might go off if a round is left in a hot chamber. If it is necessary to eject a "hot" round, stay away from it once it is out of the rifle. Though the bullet won't be dangerous if the round goes off outside the rifle, the small fragments of brass hurled out in all directions when the cartridge explodes can cause cuts or eye injury at close ranges.

If you hear a softer than normal discharge or "pop" or experience reduced recoil when you fire a round, you may have fired a bad round which left a bullet, bullet fragments, or other obstruction in the barrel. If this happens, you must stop firing immediately and get the barrel cleared: Remove the magazine, pull the charging handle back, and lock the action open. Visually inspect the chamber to be sure it's cleared. Next, place the selector in the safe position and visually inspect the bore or run a cleaning rod down it. If a jacketed bullet is wedged in the barrel, it is next to impossible to remove it. (Lead bullets can be pounded out with a cleaning rod if they must be removed in the field.)

If the rifle is necessary in combat and a bullet is stuck in the barrel, you might be successful in removing it if you remove a bullet from another cartridge and place the cartridge with its powder—but without the bullet—into the chamber and fire it. This is recommended only as a last-ditch resort and may damage the rifle.

If a cartridge is jammed in the chamber so that it cannot be fired but cannot be extracted, and if you're in combat, you're in a jam. There are several ways to remove the round, but none are too good. One is to push a cleaning rod down the barrel and push downward as you pull back on the charging handle. (This can be dangerous, since the bore is obstructed and a round could accidentally discharge.) Another method is to remove the magazine and reach through the magazine well, prying the carrier back from the barrel with a screwdriver or other tool. This may work, or it may only break the extractor and/or scar up the receiver and bolt carrier. Remember that you're working with a live round which could be set off with potentially disastrous results. You must weigh the costs before you try this, and be extremely cautious.

If you should ever have occasion to fire the AR-15 under extremely cold conditions, you may also find yourself wearing mittens or large, heavy gloves. The AR-15 trigger guard can be released to make room for mittens or bulky gloves. To release it, just take a cartridge or other small tool and push against the spring-loaded pin at the front of the guard/lower receiver hole. When the pin is depressed, rotate the trigger guard down where it can be held against the pistol grip (you may wish to keep it there by taping or wiring it in place). If you do this, remember that you have also defeated a safety device; any twig, branch, or other obstacle that gets into the trigger area might fire the rifle. To minimize this risk, keep the safety engaged until it is time to fire the rifle.

Most mishaps with the AR-15 can be prevented or minimized if you always treat the rifle as if it might fire at any moment. Never have it pointed at anything you can't afford to put a bullet through; use proper ammunition; and keep the rifle clean.

The sights on all AR-15s (except for the new M16A2 version) can be adjusted with the end of a bullet. Though this is cheap, it isn't too quick. A very useful tool for adjusting the sights is available from most mail-order companies that sell AR-15 parts. The sight adjustment tool can really speed things up. (One source: Lone Star Ordnance, P.O. Box 29404, San Antonio, TX 78229, for $8.)

To adjust "windage" or the horizontal placement of the bullet, adjust the rear sight. Most receivers have markings to show that clockwise movement changes the point of impact to the right. Depress the indent and move the drum as needed. With the full-length barrel version of the AR-15, one notch of turn will change the point of impact by one inch at 100 yards or two inches at 200 yards, etc. And yes, the sights are designed for English measurements rather than metric; this is the reason that the change in point of impact is 2.8 cm at 100 meters for each detent click.

It should be noted that with carbine versions of the AR-15 the horizontal and vertical displacement for each "click" is slightly greater since the sighting radius is shorter between the rear and front sights.

The rear sight has two range apertures on most AR-15s. One is for 0 to 300 yards and the other, marked with an "L," is the long-range aperture for 300 to 500 yards. These are just approximations, and flipping the long-range sight may or may not allow for accurate long-range shooting since bullet drop can vary considerably with the actual range and ammunition used. Nevertheless, the flip sight can be very useful.

The front sight adjusts the elevation. Be sure to stay clear of the muzzle if you're adjusting a loaded rifle. There should be an arrow and markings on the front sight to show that clockwise rotation will raise the point of impact. One click will change the point of impact one inch at one hundred yards with the standard length barrel. With Colt's A2 Sporter II, one squared-off front sight has only four click slots rather than five; that means one click will change the point of impact, up or down, one-quarter of an inch at one hundred yards.

One "trick" that can be used to sight in a rifle that has a new barrel on it is to remove the upper receiver from the lower and pull out the bolt carrier. Place the receiver/barrel on a steady rest and sight on some small object in the distance. Next, without moving the receiver/barrel, look through the barrel to see where the barrel is centered. Change the sights accordingly. Continue to zero in the sights until the sights and barrel seem to be lined up. Though this visual sight adjustment won't be precise, it will get you into the neighborhood when you finally start firing the rifle to zero in.

If you have a .22 adapter or single shot .22 adapters, they can also be used to give a rough sighting in. (The .22 CB Cap can be used for a very quiet sighting in.) Usually the zero taken at 30 yards with the .22 will be close to that of most AR-15s firing 5.56mm (.223 Remington) ammunition.

A 200- or 250-yard zero is usually the best for the AR-15 (more about battle zeros elsewhere); be sure to use the 0 to 300 range aperture when sighting in the rifle. If you are unable to sight in at this range, you can get very close to a 250-yard zero by sighting the rifle in at 30 yards. Because of the ballistic curve, the bullet is at about the same height at 30 yards as it will again be at 250 yards.

The proper sight picture is when the front post is in the center of the rear aperture vertically, with its top in the exact center horizontally. The bullet should then impact just a hair above the sight post when you're firing at the distance the rifle is zeroed at.

UNLOADING THE AR-15

Be sure the selector is in the "safe" position. Press the magazine catch and remove the magazine; pull the charging handle fully to the rear so that any round in the chamber is ejected. (Get into the habit of looking into the chamber to be sure it's empty—the rifle might fail to extract a round.) You may or may not want to lock the action open with the bolt stop. To do this, push on the bottom of the bolt release lever rather than its grooved top. The bolt release can be jarred off with an accidental hit on the rifle, so don't use the bolt release as a safety feature (i.e., if a full magazine is in the well and the bolt held open by the bolt release, if the rifle is jarred, the bolt might go forward and chamber a round so that what appeared to be a safe rifle is now ready to fire—or might even fire itself if a sensitive primer was on the top round in the magazine.)

Treat any rifle as if it were loaded.

8. Disassembly

From time to time it will be necessary to take down the rifle to lubricate it or to carry out routine maintenance. Do not take the rifle apart any more than is necessary; each time a part is removed from the lower receiver or other areas not involved in field stripping, the part's fit is not as tight when reassembled.

FIELD STRIPPING

The AR-15 is very simple to field-strip and easy to lubricate. Unlike many assault rifles, it has few small parts that can be lost during the field stripping. It's a good idea to wear safety glasses when field stripping the rifle just as it is when shooting or assembling the AR-15.

Use the exploded diagrams to locate parts mentioned in this section if you have any doubts about the piece or its location.

1. Remove the magazine from the AR-15 and check the chamber to be sure the action is closed; release the bolt carrier if it's held open so that it is closed.

2. Place the selector in the "safe" position.

3. Push out the rear take-down pin (1-35) from the selector side of the lower receiver and pull it out from the opposite side of the receiver until it locks open. This will release the upper receiver so it can be rotated downward. The front pivot pin (2-1A) may also be pushed outward if it is necessary to remove the upper receiver from the lower; on the AR-15 Sporter, the receivers are held to the front with a double screw. To remove the upper receiver from the lower on the Sporter, use two screwdrivers to unscrew the two; a drift punch may be needed to push out the inside screw from the outer one.

4. Pull the charging handle (1-14) back; the bolt carrier group (13-A, B, C, D, E) will come back with it. The bolt carrier group should be pulled back so that it can be grasped and removed. After the carrier group is removed, the charging handle can be removed by pulling it back and downward through the slot in the key channel.

5. The firing pin retaining pin (1-1) can be removed by punching it out with a tip of a small tool. Push from the ejection port side of the carrier.

6. Once the retaining pin is removed, the firing pin (1-2) can be removed by tilting the carrier up so the bolt faces upward and the firing pin can fall free through the rear of the carrier.

7. Twist the bolt assembly (1-4) so that the cam pin (1-3) is clear of the bolt key (1-13C). The cam can now be lifted free and removed.

8. With the cam pin removed, pull the bolt assembly (1-4) from the front of the bolt carrier. Care should be taken in disassembling the bolt as a number of easily lost parts will be freed.

9. The extractor spring (1-7) is under compression, so be careful not to let the extractor (1-6) get away when it is released. While holding the extractor in place, use a small wire or drift punch to push the extractor pin (1-5) out of the bolt. Carefully remove the extractor, its spring, and the small nylon plug that is often nestled in the spring.

10. The ejector (1-9) and its spring (1-10) can be removed by drifting out the pin (1-8) that retains it. The spring is under tension so be careful not to let it get away. Placing an empty brass into the bolt face and using it to hold the ejector in place can make the removal of the ejector pin easier.

11. It is seldom necessary to remove the handguards (3-1, 2) unless they need to be replaced or the weapon is full of mud. Although in theory the handguards are easily removed by pushing the

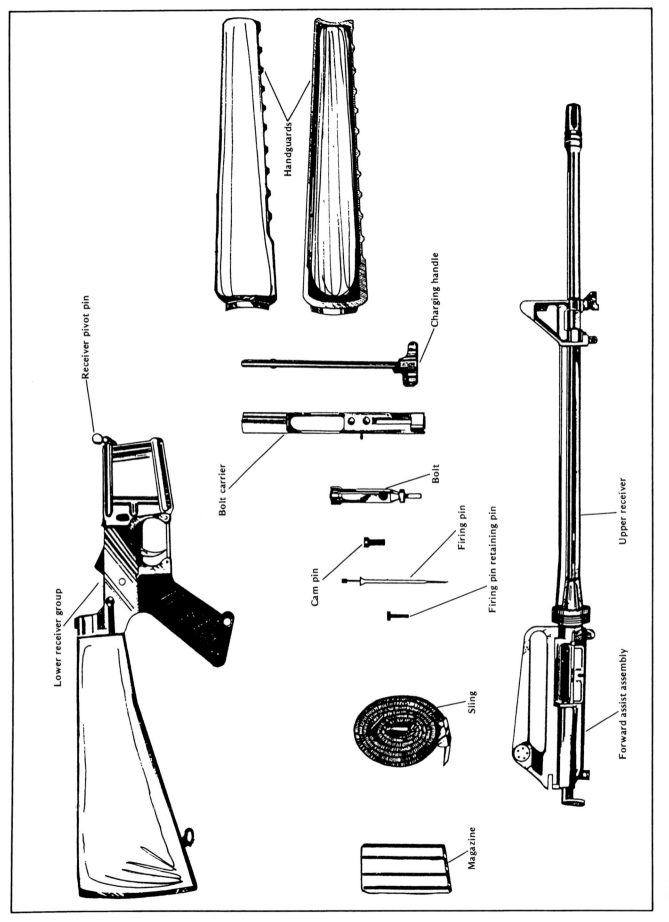

A field-stipped AR-15

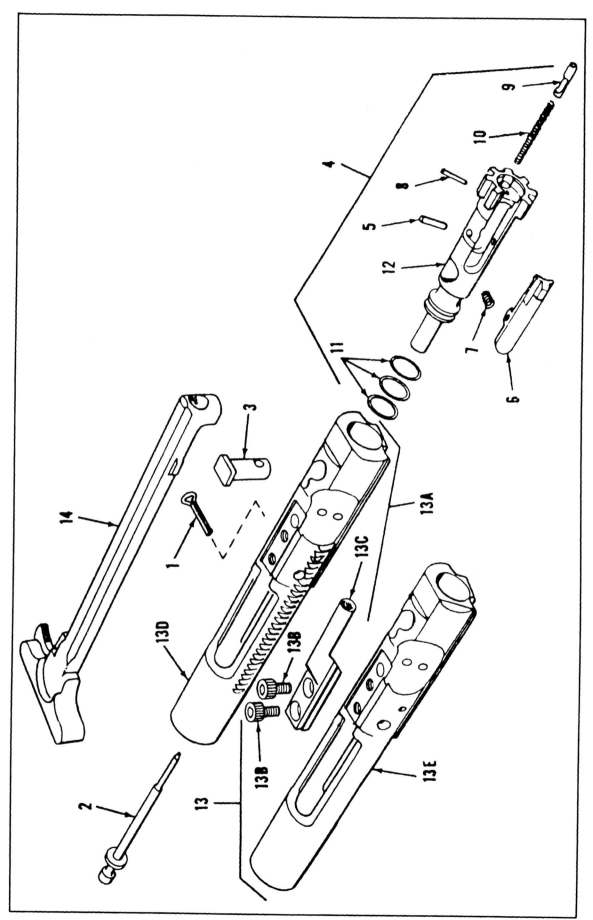

DIAGRAM 1 — BOLT CARRIER GROUP

1-1: Firing pin retaining pin; 1-2: firing pin; 1-3: bolt cam pin; 1-4: bolt assembly; 1-5: extractor pin; 1-6: extractor; 1-7: extractor spring; 1-8: ejector pin; 1-9: ejector; 1-10: ejector spring; 1-11: gas rings (3); 1-12: bolt; 1-13: bolt carrier assembly (without forward assist grooves)); 1-13A: bolt carrier assembly (forward assist style); 1-13B: key bolts (2); 1-13C: key; 1-13D: bolt carrier (with forward assist grooves); 1-13E: bolt carrier (without forward assist grooves); 1-14: charging handle assembly.

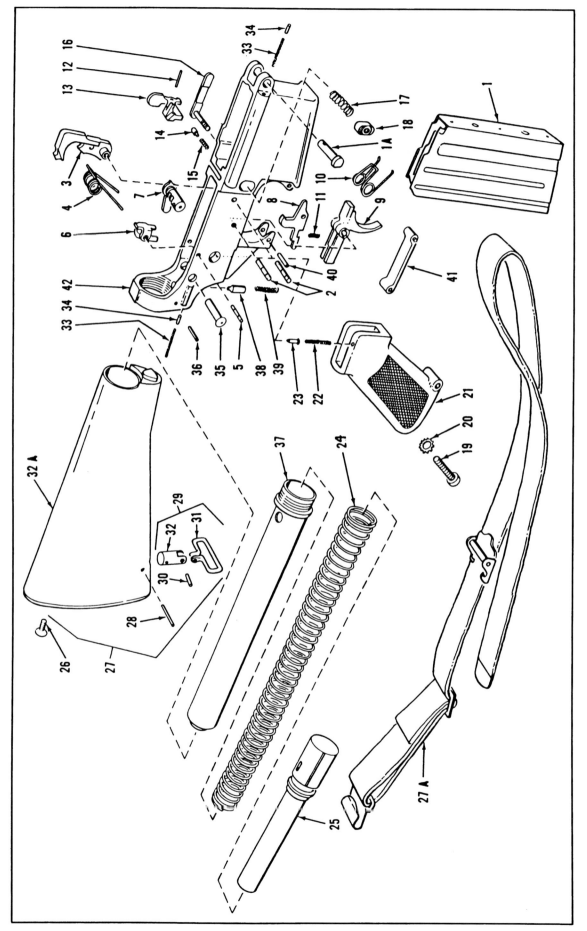

DIAGRAM 2 — LOWER RECEIVER GROUP

2-1: Magazine; 2-1A: front push pin (pivot pin—dual screws on some models); 2-2: hammer/trigger pins (2); 2-3: hammer; 2-4: hammer spring; 2-5: auto sear pin; 2-6: automatic sear; 2-7: selector leve'; 2-8: disconnector; 2-9: trigger; 2-10: trigger spring; 2-11: disconnector spring; 2-12: bolt catch pin; 2-13: bolt catch; 2-14: bolt catch plunger; 2-15: bolt catch spring; 2-16: magazine catch; 2-17: magazine catch spring; 2-18: magazine release button; 2-19: pistol grip screw; 2-20: pistol grip lock washer; 2-21: pistol grip; 2-22: safety detent spring (same as ejector spring); 2-23: selector lever (safety) detent; 2-24: buffer spring; 2-25: buffer; 2-26: top stock screw; 2-27: stock assembly; 2-27A: sling; 2-28: rear swivel pin (old-style stocks only ; 2-29: rear swivel assembly (old-style stocks only); 2-30: rear swivel pin (old-style stocks only); 2-31: rear swivel (old-style stocks only); 2-32: swivel stud; 2-32A: stock; 2-33: rear push pin detent spring; 2-34: rear push pin detent; 2-35: rear push pin (takedown pin); 2-36: buffer pin (military models only); 2-37: buffer tube (lower receiver extension); 2-38: buffer retainer; 2-39: buffer retainer spring; 2-40: trigger guard roll pin; 2-41: trigger guard and its front spring and pin; 2-42: lower receiver.

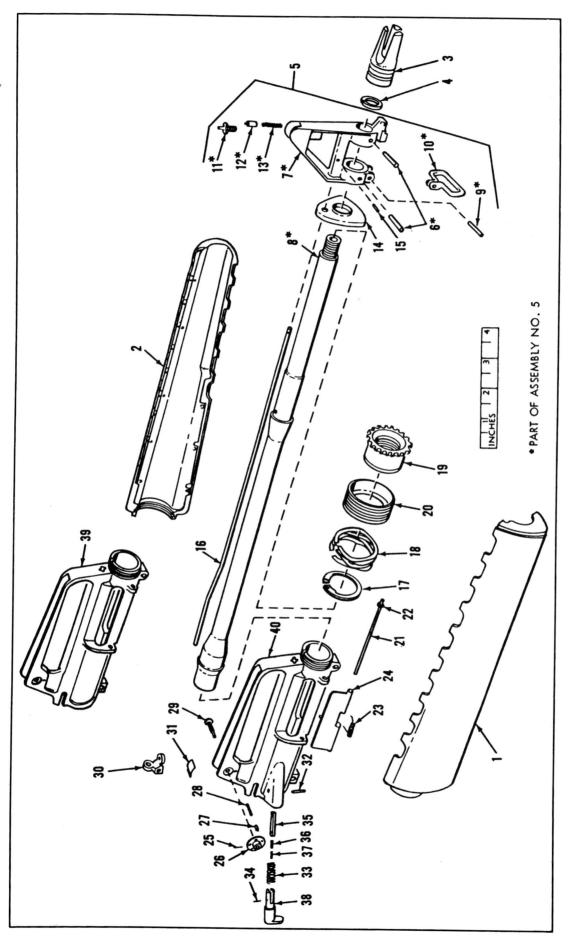

DIAGRAM 3 – UPPER RECEIVER GROUP

3-1: Handguard half (old style); 3-2: handguard half (old style); 3-3: flash suppressor; 3-4: flash suppressor lock washer; 3-5: barrel and sight assembly; 3-6: front sight; 3-7: front sight; 3-8: barrel assembly; 3-9: front swivel pin; 3-10: front swivel; 3-11: front swivel pin; 3-12: front sight detent; 3-13: front sight detent spring; 3-14: handguard cap; 3-15: gas tube pin; 3-16: gas tube; 3-17: snap ring; 3-18: slip ring spring; 3-19: barrel nut; 3-20: slip ring; 3-21: ejection port cover pin; 3-22: ejection port cover retaining ring; 3-23: ejection port cover spring; 3-24: ejection port cover; 3-25: rear sight windage drum pin; 3-26: rear sight windage drum; 3-27: rear sight detent; 3-28: rear sight detent spring; 3-29: rear sight screw; 3-30: rear sight; 3-31: rear sight spring; 3-32: forward assist pin; 3-33: forward assist plunger spring; 3-34: pawl pivot pin; 3-35: forward assist pawl; 3-36: pawl detent; 3-37: pawl detent spring; 3-38: forward assist; 3-39: upper receiver (without forward assist well); 3-40: upper receiver (with forward assist well).

* PART OF ASSEMBLY NO. 5

INCHES 1 2 3 4

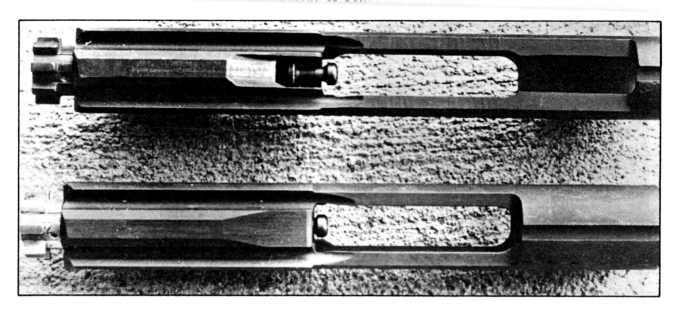

Top: semiauto bolt carrier. Bottom: selective-fire carrier.

weld ring (3-20) toward the receiver, in reality there is often enough tension in the weld spring (1-18) to make this hard for most people to do.

One solution is to use a screwdriver or other tool to carefully pry the ring away from one half of the handguard until it can be wiggled free. Repeat the procedure with the other half of the hand-guard. This will have to be repeated to get the handguards back on, if it was necessary to pry them off. Be careful not to scar up the finish on the rifle when using a tool as a lever on the weld ring.

12. To remove the buffer (2-5) and its spring (2-25), it is necessary to depress the buffer retainer plunger (2-38) while restraining the buffer and its spring. Be sure the hammer is in its cocked position before trying to remove the buffer from its tube. Once the buffer is out, the spring can be wiggled out without holding the plunger down.

Assembly of the rifle is basically a reversal of all the above procedures. Be sure the cam pin and firing pin retaining pin are in place as the rifle would be dangerous to fire without them.

TOTAL DISASSEMBLY

All other disassembly of the AR-15 or any major work done on the rifle should be done by a competent gunsmith. Work on your rifle *only* if you are sure of your abilities to do so. Probably more rifles are ruined by amateur work than by accidental breakage. Don't be too proud to get someone else to do work that you can't do. Such pride can be expensive or even dangerous in the long run. Do

not work on your rifle if you are not sure of what you are doing!

The only time that you might attempt the following disassembly (unless you are experienced with working on firearms) would be in a survival or combat situation where you might prefer the risk of damaging your gun than being without a working weapon.

If you must disassemble the rifle, do only the following steps that are necessary. You should skip any that are not necessary to carry out the work that needs to be done. Consult the section of this book on assembling an AR-15 to reassemble the rifle if you run into any problems.

1. The three gas rings (1-11) should be left in place unless they are damaged. New rings are placed on the bolt by gently spreading each one enough to get it in place. Don't line the spaces up on the rings.

2. The charging handle latch and its spring should be left on the charging handle (1-14) when possible. The roll pin holding them can be drifted out of place if necessary.

3. The key (1-13C) and its bolts (1-13B) should be left in place unless they absolutely need to be replaced. It may be necessary to grind off the metal holding the bolts in place as they are peened on most carriers. A hex wrench is necessary to re move the two bolts.

4. The flash suppressor (3-3) screws off the barrel. Be careful not to lose the lock washer (3-4). The flash suppressor is most easily removed by clamping the barrel into place.

5. The front sight assembly (3-5) can be removed by drifting out the two pins (3-6) from the base of the assembly. The pins must be pushed out toward the ejection port side of the rifle. The base, along with the gas tube (3-16), can be moved off the barrel from the muzzle end of the barrel.

6. The front swivel (3-10) can be removed by drifting out the pin holding it. On some AR-15s this is a rivet that must be ground off.

7. The front sight post (3-11), sight detent (3-12), and their spring (3-13) can be released by depressing the detent and unscrewing the front sight post. A front/rear sight tool is useful for doing this.

8. If the front sight has been removed, the gas tube (3-16) can be removed by drifting out its pin (3-15). The gas tube can also be removed without removing the sight base by drifting out the gas tube pin and pushing the tube to the rear of the upper receiver, pulling the tube slightly to the side of the front sight base, and pulling it out of the receiver toward the muzzle of the barrel.

9. With the gas tube removed, the barrel nut (3-19) can be unscrewed to remove the barrel. An armorer's wrench is necessary to do this. The barrel can be pulled straight out of the receiver once the nut is removed (it may be necessary to wiggle the barrel a bit to make it creep out).

10. The handguard snap ring (3-17) along with the weld ring (3-18) and slip ring (3-20) will stay on the barrel nut when it is removed. If you wish to remove them, use a pair of needle-nosed pliers to pull the snap ring free of its groove in the barrel nut.

11. If the barrel is off, the ejection port dust cover (3-24) along with its spring (3-23) and its pin (3-21) can be removed by sliding the pin toward the barrel side of the receiver. If the barrel is left on the upper receiver, then the "C" ring (3-22) can be popped off the pin, and the pin slipped out toward the rear of the receiver thereby freeing the port cover and its spring. (This is a hassle with receivers that have a bolt assist, but it is possible to do on all AR-15 rifles.)

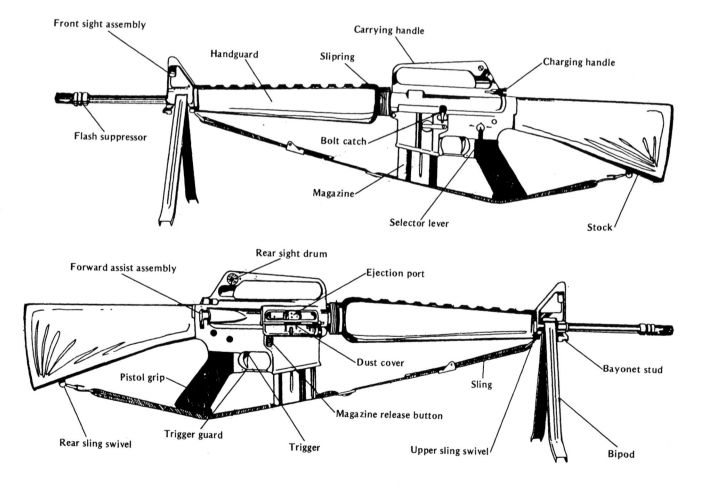

Parts identification diagram

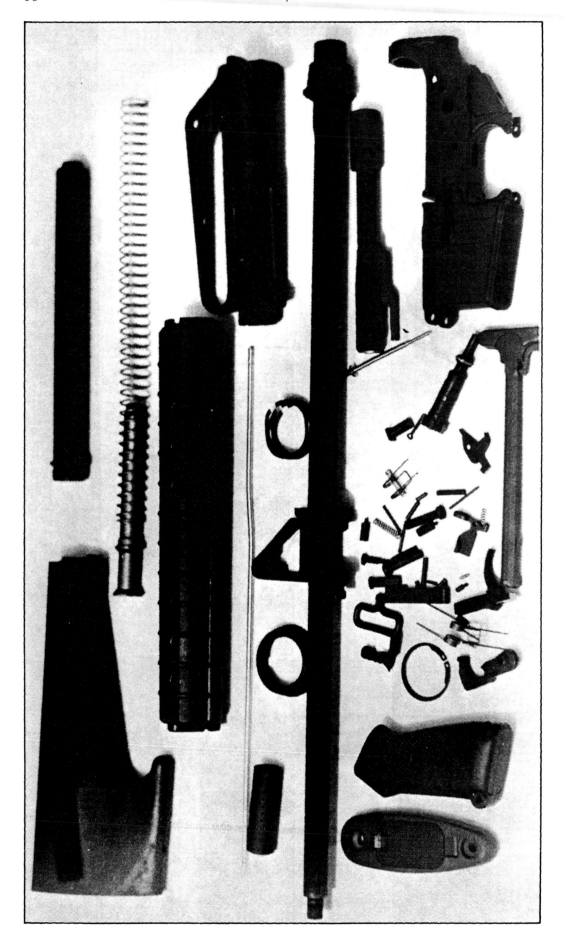

The AR-15 as it appears prior to assembly

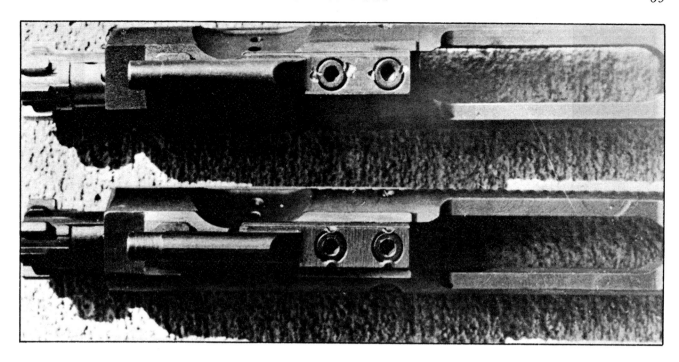

Carrier key bolts can be peened into place.

12. The rear sight windage drum (3-26) can be removed by drifting out the windage drum pin (3-24) and removing the drum, detent (3-27), and detent spring (3-28). The spring is under tension so be careful when freeing it.

13. The rear sight (3-30) can be removed by unscrewing the rear sight screw (3-29). This will free the rear sight spring (3-31).

14. If the rifle has a forward assist, the assembly can be removed by drifting out the pin (3-32) that holds the assembly. Generally it should be drifted out from the top of the receiver toward the lower part of the receiver.

15. The forward assist assembly can be disassembled by drifting out the pawl pivot pin (3-34).

16. Remove the pistol grip (2-21) and its screw lock ring (2-20) by unscrewing its screw (2-19). This will also release the selector detent (2-23) and its spring (2-22).

17. Once the pistol grip is off, the selector (2-7) can be removed from the bolt release side of the receiver. (The hammer should be cocked to do this.)

18. The hammer (2-3) and its spring (2-4) can be removed by drifting out their pin (2-2). The spring can be removed by gently pulling one side off the hub. Remember how the spring is positioned on the hammer if you have to remove it.

19. The trigger (2-9), trigger spring (2-10), disconnector (2-8), and the disconnector spring (2-11) can be removed by drifting out the pin (2-2) holding them. The spring can be removed by gently pulling one side off the hub. Remember how the spring is positioned on the trigger.

20. If the rifle has one, the auto sear (2-6) can be removed by drifting out its pin (2-5).

21. Use a drift punch to remove the bolt release pin (2-12). This will free the bolt release (2-13), its plunger (2-14), and its spring (2-15).

22. Push the magazine release button (2-18) down as far as possible with a small tool and unscrew the magazine catch (2-16). This will free the two parts along with their spring (2-17).

23. The front push pin (2-1A) can be removed on most rifles by pulling the pin all the way to its release position, then inserting a small tool or wire from the barrel side of the pin through the hole in it. This will depress the detent (2-34) and its spring (2-33), allowing you to pull the pin out. Be careful not to lose the spring and detent as they are under tension.

If the push pin does not have a hole in it through which the wire can be inserted, it will be necessary to use an L-shaped tool which is inserted down the pin groove and used to depress the detent so the pin can be pulled free. This tool can be made by grinding a small L-wrench down to a flat bar on its short end.

24. The stock (2-32A) can be removed by unscrewing the screw (2-26) at its rear. This will also

release the rear pin detent and its spring which in turn will free the rear push pin (2-35).

25. The buffer tube (2-37) can be unscrewed on most models of the AR-15. On some rifles (especially military weapons) a pin (2-36) may have to be drifted out to allow the tube to be unscrewed. If the tube has Loctite or similar material on it, it may be necessary to dissolve it with acetone.

26. On old style rifles (without the trap door on the stock), the rear swivel (2-31, 2-32) can be removed and disassembled by drifting out its pins (2-28, 2-30).

27. On rifles with a trap-door stock, the lower screw can be removed to separate the door assembly from the stock and to free the rear swivel.

28. On the telescoping stock, the rear of the stock can be removed from the buffer tube by levering up the spring-loaded bolt that holds its release handle down. This will allow the stock to be slid off the buffer.

29. To remove the trigger guard (2-41), use a small tool to depress the front pin (which is spring loaded) so that it releases the front of the guard. Next, drift out the rear roll pin (2-40).

For assembly tips, see the section on building an AR-15 rifle.

DISASSEMBLY OF MAGAZINES

The magazines should be disassembled occasionally for cleaning and lubrication. On metal magazines, the metal base on the bottom of the magazine slides out. Some magazines require a small tool to release the plate, which may be held in place by pressure from the spring.

Newer plastic magazines generally have a catch somewhere on their side. Locate it and gently push it in with a screwdriver or similar tool. When removing the bottom plate from a plastic magazine, be careful not to over flex it as it might become deformed or even crack.

The spring is under pressure and will pop part way out after the floor plate is removed. Jiggle the spring the rest of the way out. The follower will come out with the spring. Do *not* remove the follower from the spring. If the spring comes loose from the follower, quickly note the way the spring is aligned in the magazine and put the follower back on the spring before doing anything else. If the spring is placed on the follower backward, the magazine will not feed properly!

Carefully wipe all dirt from the magazine and very lightly lubricate it. Too much oil will attract dirt.

When reassembling the magazine, jiggle the spring back into place after placing the follower so that its front faces the front of the magazine.

Slide the base plate back into place. Be sure the printing is facing so that you can see it on the outside of the bottom of the magazine. Check to make sure the base plate is secure and won't come loose.

9. Building Your Own Rifle

Why build an AR-15? For one thing, building your own rifle is a real ego trip. But there are a lot of other good reasons. You can create one of the AR-15 variations that is not available on the market; in effect, a custom rifle suited to your needs.

Owners of automatic versions of the AR-15 can benefit from a little do-it-yourself work. Since the lower receiver carriers the automatic fire mechanism and is the part that is registered, building a few upper receiver/barrel assemblies can make it possible for the holder of the automatic weapon permit to "own" the equivalent of several automatic weapons while paying the tax and doing the paper work for only one auto weapon.

Suppose you own an automatic AR-15 with a regular length barrel. By building a few upper assemblies, you could also have a short-barrel carbine, a heavy-barrel machine gun, a long-barreled sniper rifle, and a submachine gun in .22 LR (via a separate barrel/.22 adapter or just a .22 adapter). If you go on to buy a .45 ACP or a 9mm upper conversion you could also have a submachine gun in these calibers! Six automatic weapons with the tax and paperwork of one.

With a lower receiver with push pins, you can do the one-lower-with-a-number-of-uppers trick with a semiauto action, too, and realize quite a savings over what the equivalent arsenal would cost with any other rifle.

Since you'll only shoot one rifle at a time (I hope), having one lower to a number of uppers isn't bad except for the looks of your gun rack. As mentioned elsewhere, having a family of rifles—a weapon system—that have identical "controls" makes for considerable savings in training time when using a new weapon, savings if you purchase spare parts, and for a lot less confusion when you switch from one rifle to another.

If you build your own rifle, you will also be able to do a lot of the repair work on it when you have to (most modern gunsmithing work consists of replacing parts—something you can do after building your own rifle). Knowing how to repair your weapon can save a lot of money and could be a life-saver during a battle.

You may realize a slight savings by building a standard AR-15 rifle, but this should not be your sole purpose for building a rifle since the savings will be small, and—if your time is worth anything—you'll come out about even with what it would have cost you to get a used Colt AR-15.

You will save money if you want to create a special type of AR-15 or modify your rifle with a lot of accessories or replacement parts, such as a new pistol grip/handguards, stock, flash suppressor, heavy barrel, modified upper receiver, or upper receiver with forward assist. In such a case, building your own rifle allows you to buy the parts that you want rather than buying a stock rifle and then buying the accessories and replacement parts for it. You will end up with an inexpensive rifle rather than an inexpensive rifle *plus* a pile of expensive parts that will probably never be needed even as replacement parts.

Building your own rifle can also enable you to obtain a special version of the AR-15 that is not available on the market. Examples of what you could build that is not currently on the market include a rifle with a stock an extra inch or three-quarters of an inch long; a rifle with a regular stock with a short 16-inch carbine barrel; a heavy-barreled sniper rifle; a rifle with a scope mount (and no carrying handle or sights); etc.

It should be noted that at the time of this writing, Quality Parts Co. (P.O. Box 6659, Woodfords/Portland, ME 04101, 207/775-1744) will assemble a rifle from parts that you order from them. Rock Island Armory (420 West Main St., Geneseo, IL 61254, 309/944-5739) makes a number of different, fully assembled AR-15s, including short barreled, automatic versions that may not be available with Colt-only parts; they will assemble a rifle to your specifications as well.

Is it legal to build a firearm without a permit? Usually. The U.S. government does not place limits on the firearms that people make for themselves provided they are not outside the legal limits of firearms laws, i.e., the minimum length of the barrel is 16 inches, the minimum overall length is 26 inches, and the weapon is semiautomatic. You would be wise to check into state and local laws. Call the city and state attorneys to find out the laws. Often police departments are completely ignorant of the laws.

You might run into some problems if it appears as if you are building rifles to make money. This is because the Gun Control Act of 1968 was very poorly written so that the term "dealer" is never well defined; nor does the act really come to grips with what a firearms manufacturer actually is.

You might be considered a dealer by a BATF agent if you make a few dollars by selling the one gun you have assembled. Conversely, at the time of this writing, you probably would not be considered a manufacturer or dealer if you assembled one or two rifles in a year and traded them or sold them at a slight profit. (This is only if you make a rifle complete with lower receiver. You probably could break the rifle into parts and sell the lower receiver at cost while still making a profit on the other parts.)

You can write to the Department of the Treasury, BATF, Washington, DC 20226, to get an opinion if you have made more than one AR-15 that you wish to sell. BATF people are generally very willing to help and will probably give you a written okay to sell your rifles. If they don't, a Federal Firearms License (FFL) can be obtained for just $10, and then you can sell as many guns as you wish or buy guns through the mail.

The final consideration is how hard is it to build an AR-15? Because the AR-15 was designed to take advantage of modern manufacturing techniques that make it possible for machines to easily make parts that stay within the specifications of the rifle, it is relatively easy to build an AR-15 or any of its variants from parts. The AR-15 was designed to allow normal manufacturing tolerances without degrading the performance of the rifle when parts are put together without extensive fitting.

Though many people are a little leery of tackling such a project, in fact anyone who can chew gum and work a screwdriver at the same time can probably assemble an AR-15 from a pile of parts. At the same time, it should be stressed that a rifle that doesn't function properly is very, very dangerous. Therefore, if you have any doubts about tackling this project, try to find someone to help you with it. Do not proceed with building a rifle if you have

The author is seen here with a do-it-yourself rifle.

some doubt about whether or not you can do it.

The time it will take you to build an AR-15 from parts will vary according to your abilities and the extra work required if a few parts have to be hand fitted together (which is pretty rare except for cleaning up the finish on the receiver—more about that later). The main thing is to take your time and do it right. If you don't have the patience to spend several evenings working at the step-by-step process, forget it. You can't rush through it or you will end up with a piece of junk. In general, you will probably need about a week of evenings to get the job done, though some people can assemble a rifle in a couple of hours if they have all the right parts and everything fits the way it should.

If you want to build a rifle, your first task is to collect the parts you need. Try to get all the parts before you start building; nothing is as frustrating as getting halfway through the building procedure and then having to wait a week for a part you ordered.

Of course the best parts to use are brand-new Colt parts. But the price for Colt parts is usually at a premium, and often the "genuine Colt" parts that some dealers sell are not made by Colt. Even those parts that Colt sells aren't all manufactured in-house; some of the parts are made by other manufacturers for Colt. Also, Colt does not make a lot of the accessories or parts that you may want to use. You may have to buy non-Colt parts to build the rifle you want.

So you will find yourself, for one reason or another, with non-Colt parts. The question is, how good are non-Colt parts? Most of the ones I've used in building my rifles have worked very well. Even those that were poorly cast or distorted by being removed too quickly from a mold usually work well—even if they look like they would never do so—provided they fit in the proper position.

If you're buying your parts "in person" from someone (say at a gun show—not always the most reliable of places), try to look at the part under a magnifying glass to see whether it has sharp or machined-looking edges and surfaces. That is the way a part should look. If the part has less definition on its edges and maybe even looks slightly melted or has bubbles on its flat surfaces, then avoid purchasing it.

Although it would be wise to stick to factory parts, it is sometimes possible to "roll your own" parts. This is especially true with the roll pins, springs, push pins, and the like. More than once while assembling my own rifles I have had to make a small roll pin substitute or wind my own spring. (An excellent spring-winder tool is available from Brookstone, 127 Vose Farm Rd., Peterborough, NH 03458, 603/924-9511, for just $13. Brookstone also offers an assortment kit of foot-long spring wire for $9 to get you started.)

It is possible to leave some parts out when you build a rifle, although it isn't advisable. The rifle will operate without the bolt catch, dust cover, forward assist assembly, and the front and rear sights (if you're planning on using a scope). Doing without these parts might allow you to build a rifle for a smaller investment. Then you can gradually add parts as you get the cash and decide what you want to add.

If you are trying to save money, a better way than leaving off parts is to watch for bargains that are offered from time to time by most mail-order companies. These offers allow you to buy the parts for the upper receiver, lower receiver, or even all the parts for the whole rifle at a considerably smaller cost than you would pay if you purchased each part separately.

One part that sells at a premium is the Colt telescoping "shorty" stock. These are good, but the Colt stock can cost two to three times as much as the stocks made by SGW or other companies. The other stocks are practically as good. Unless you really feel a need for the Colt stock, buy one of the less expensive stocks.

Most parts for the automatic version of the AR-15 will fit in the semiauto Sporter version and vice versa. There are some spots, however, where you can't use whatever is handy. Two parts that you would do well to match up are the upper and lower receivers. The Colt Sporter semiauto upper and lower receivers don't match up with the military-style receivers. Use a pair of one or the other, but do not mix them unless you really have to. When possible buy them as a pair from the same source.

It *is* possible to mix the receiver halves. You can even put a military upper or lower receiver with the opposite number from the Sporter if you use an offset adapter pin rather than a front push pin. These are available from a number of dealers. Just be sure to get the right one: A different one is needed for the Sporter upper/military lower than is needed for the military upper/Sporter lower.

Frankly, the only reason to go with the Sporter upper or lower would be if you can get one at a

Shown here and on the previous page are some versions of the AR-15 the author has built.

super-low price or if you don't want the forward assist on the upper receiver. The forward assist is not really necessary, although it looks more "military"—an element of some importance to Walter Mitty types. The forward assist is, however, handy when you want to quietly chamber a round or when you're using the gun in an extremely cold environment.

The military upper receiver does have a real plus: the forward push pin. While the Sporter AR-15 has a double screw on the front of the receiver halves, the military version has a push pin similar to the rear one. The rear pin is used as it is on the Sporter to break the action open when it's time to clean the rifle. The front pin allows you to quickly take the gun apart when you wish to store or transport it, or when you need to place a new upper receiver barrel assembly on the original lower receiver quickly.

Two other parts that must be either Sporter or military are the disconnector and the trigger. There's no way to mix the two since the spring well in the trigger is in a different spot for each type. Either go with the AR-15 Sporter (usually listed as "AR-15" parts in the mail-order catalogs) or the AR-15 military parts (generally called "M16" parts).

The M16 parts kits are usually cheaper and work just as well as the Sporter parts. Just be sure to make the modification (described in the assembly steps) to the disconnector as soon as the parts kit comes in.

If you order a parts kit that does not specify which type of parts you'll be getting, chances are that it will contain the military automatic rifle parts *without* the auto sear. These parts work well and are generally the better buy. Since the automatic sear is missing from the kit, you should not run into any problems with the BATF in regard to having an illegal automatic weapon provided you make the one small modification to the disconnector outlined below.

Where should you get the parts? You can pick them up at gun shows, but be very careful if you go this route. A lot of good parts, which have very little wear, are sold as government surplus There are also a lot of worn parts stripped from old M16 rifles available, and they are more apt to turn up at a show than on the shelves of a reputable parts dealer. If you buy parts at a gun show, be sure to look them over carefully. Since it is

possible to reblue or phosphate old parts, parts may not necessarily be new just because they have a new finish.

Personally, I prefer to buy parts from mail-order companies. Though it occasionally takes a little longer to get the parts, they are almost always of good quality, and the companies will replace a defective part (provided you discover the defect within several days of receiving the parts).

There are a number of companies that deal in rifle parts and accessories for the AR-15. The ones I've had the best luck with are: Lone Star Ordnance (P.O. Box 29404, San Antonio, TX 78229, 512/681-9280); L.L. Baston Co. (Box 1995, El Dorado, AR 71730, 800/643-1564); M.A., Inc. (Box 5383, Shreveport, LA 71106, 318/797-3491); Numrich Arms (West Hurley, NY 12491, 914/679-2417); Quality Parts Co. (P.O. Box 6659, Woodfords/Portland, ME 04101, 207/775-1744); Rock Island Armory (420 West Main St., Geneseo, IL 61254, 309/944-5739); SGW (624 Old Pacific Hwy., S.E., Olympia, WA 98503); Sherwood International (18714 Parthenia St., Northridge, CA 91324, 800/423-5237); and Springfield Armory (420 West Main St., Geneseo, IL 61254, 309/944-5138).

I have used parts from all the companies listed and have purposely mixed parts when building rifles to see if any problems would occur. I have found no problems created by poor manufacture except for one part that looked like it had dropped from a mold while it was still hot (it was a little bent out of shape).

All the companies will accept personal checks (except for M.A., Inc.), but for the very fastest service, you should send a money order or use a Visa/MasterCharge card. (You can also send cash; the postal service frowns on this, but I've never had any problems doing it.)

M.A., Inc., gives the fastest service; they'll get the parts shipped within forty-eight hours upon receipt of your order, and the parts are sent via UPS. Usually, the parts can get to you within a week. The company does not back order parts, which you may find an advantage, depending on how much of a hurry you're in.

The other companies generally can get the parts to you within two weeks *if* you do not use a personal check. (A personal check takes longer since it has to be processed by a company's local bank to be sure the check doesn't bounce.)

Most of the time the companies will ship via

The author has also built these three variations of the AR-15.

overnight carriers, C.O.D., or the like *if* you are willing to pay the extra charges.

The only part that can't be ordered without a Federal Firearms License is the lower receiver. Your local gunsmith or gun store should be able to order one for you for a small fee. Try to get both the upper and lower receiver at the same time so you'll get a matching set.

When you build a rifle yourself, you don't need any special permit from the federal government but you should check local government regulations. In general, if you make the rifle for your own use, it's legal to assemble it.

PARTS LIST

Here's a parts list of what you'll need to complete an AR-15 (it's quicker and easier to buy parts sets, so that's all that's listed. Use the exploded diagram if you're going for separate parts):

1. Lower receiver. The military style with forward push pin is usually best.

2. Upper receiver. The military style with push pins and forward assist is usually best.

3. Forward assist assembly (if you have a military-style upper receiver). Two types are available: the round handled and the tear-drop shaped handle. The tear-drop is considered better by many, but you won't go too wrong by just getting the least expensive style.

4. Lower receiver parts set.

5. Complete bolt. Avoid the chromed bolt since the chrome generally flakes off.

6. Bolt carrier with screws and key. Avoid the all-chromed bolt carrier because it wears out the inside of the upper receiver. The carrier should be chromed on the inside where the bolt will ride. Bolt carriers for use with the forward assist need to have slots cut down the ejection port side for the assist claw.

7. Charging handle.

8. Firing pin. This part should be chromed.

9. Firing pin retaining pin.

10. Bolt cam pin.

11. Buffer tube. It is not needed with a "shorty" telescoping stock.

12. Buffer spring. This part comes with most shorty stock kits.

13. Buffer. The buffer usually comes with shorty stock kits; a special one is needed with the telescoping stock.

14. Stock end plate. It is not needed with shorty stock. This part may come with the stock.

15. Rear swivel. This part may come with the stock and is not needed for the shorty stock.

16. Stock. Several different styles are available including the shorty telescoping stock.

17. Pistol grip. A number of styles are available, but the trap door style is probably the best.

18. Pistol grip screw and lock washer.

19. Barrel handguards. Several styles are available, but handguards with aluminum liners are the best. Most shooters prefer round handguards.

20. Rear sight assembly. You may wish to buy the new style with finger adjustable windage compensator.

21. Front sight assembly. This part may already be with the barrel.

22. Gas tube and roll pin.

23. Barrel. A number of lengths are available as well as heavy barrels and stainless steel barrels. A chrome lining, if available, is a good idea. Try to get a barrel with the front sight assembly mounted and barrel nut in place; avoid stainless steel barrels if possible for combat rifles since they quickly overheat.

24. Barrel snap ring.

25. Barrel weld spring.

26. Barrel slip ring, also known as the Delta ring. Get a tapered ring if possible.

27. Flash suppressor. The government style is cheapest and works well, though many prefer those with built-in compensators.

28. Flash suppressor lock ring.

29. Front swivel.

30. Front swivel roll pin.

TOOLS

You can build an AR-15 with just a few tools that would normally be in any household shop, but there are a few that will make your job a lot easier.

One handy tool is an M16 armorer's wrench, which allows you to tighten or loosen the barrel nut, the flash suppressor, some styles of buffer tubes, and a few other odds and ends. This tool is usually available from the same company from which you buy your parts.

Barrel vise jaw blocks are very handy, the best being made of aluminum so that they do not scratch the barrel. They work with a regular shop vise and hold the barrel in whatever position is needed. When you're working on your rifle, the blocks are almost as handy as having a third hand to help you hold things. They are available from a number of companies which sell AR-15 parts.

The AR-15 does not have a lot of screws, though there are a few. When you use a regular screwdriver on a gun, you generally mess up the slot because gun screws are more shallow and thinner than regular screws. If you want to put together a rifle that looks like it was professionally done (rather than something a trained ape assembled), you would do well to get a set of good gun screwdrivers.

My favorite set is the Chapman Gun Screwdriver Set available from Parellex (1285 Mark Street, Bensenville, IL 60106, 800/323-3233) for $20. This set includes a number of blades, as well as a driver and a ratchet extension. (In addition to being handy for assembling your rifle, you'll find yourself using it around the house all the time.)

A front/rear sight tool can facilitate matters when you are assembling the front sight and when you're zeroing in the rifle. It depresses the detent on the front and rear sights and allows you to make large adjustments quickly. With the new M16-A2 or AR-15s modified with a hand-adjustable windage knob, this tool may not be as useful.

A set of chamber gauges is practically essential if you are buying a barrel, used gun, or reloading ammunition for your rifle. A nice set of these is available from Quality Parts, as well as a number of other companies. Gauges currently cost $20 each, and a set of three ($60) is ideal.

The gauges can be used to determine whether or not the chamber is the proper size. Though barrel chambers purchased from the companies recommended above will usually be the correct size, those bought from other sources may not be. If you frequent gun shows or have some doubts about a barrel that seems too cheap to be true, you would be wise to check it with a set of gauges.

When a rifle is fired a number of times, the head space may change. This normally is not a problem *unless* you are using hot loads. The chamber/bolt may then become so battered that the chamber actually stretches and the bolt lock-up lugs become deformed. In such a case, excessive headspace may result. This means that the brass in the chamber will not be supported by the chamber walls when a round is fired. Most likely the case will burst, and then gases will escape around the bolt and explode back into the receiver and magazine well. If you are lucky, you will only ruin your rifle instead of a bystander or yourself! Headspace gauges can be used to avoid such problems.

GAUGES

There are three types of gauges: "Go," "No-go," and "Field" gauges. To use the gauges, it is usually wise to remove the ejector and extractor. (If you are in a hurry, you can leave them in place *if* your gauges have a cut-out for the ejector and a groove for the extractor. Be sure to line up the ejector hole on the gauge with the ejector when you drop the gauge into the chamber. Be gentle, or you will be replacing the extractor or ejector after breaking them!)

If you are checking out a barrel that is not mounted in a receiver, screw it onto a receiver and place a bolt and bolt carrier in the receiver. (You could simply try fitting a bolt to the barrel and turning the lugs into their positions in the barrel extension by hand, but you really need to know what you are doing.) If you are checking the chamber on an assembled AR-15, it is best to pull out the rear push pin and swivel the top receiver down.

The rifle or receiver/barrel is held with the muzzle pointing down. The gauge is dropped into the chamber; the gauge is lined up with a small tool if you have not removed the ejector from the bolt; and the bolt and bolt carrier are pushed toward the chamber. By watching the bolt from the bottom of the upper receiver, you can tell whether or not it turns and locks into position. Do not force the bolt. Be sure to use gentle pressure to see whether or not the bolt will lock with the gauge in place.

With the go gauge, the bolt should close and lock with the gauge in the chamber. This means the chamber will accommodate any properly loaded ammo, down to the minimum dimensions for the .223/5.56mm, and that the rounds will chamber easily (provided the chamber is reasonably clean when you are shooting).

If the action does not close on the go gauge, the chamber has not been set to the correct depth or the bolt is oversized. (Changing bolts may help, but the problem will generally be in the chamber size.) Though the barrel might accept some cartridges, the chance that it will not chamber some rounds would be great.

A rifle which won't accept a go gauge may also be prone to slam fires when the chamber becomes somewhat fouled; cartridges hang up in the chamber and the floating firing pin then hits the primer with extra force, firing the cartridge.

Purchasing a barrel or rifle which does not accept the go gauge should obviously be avoided whenever possible. Such a barrel or rifle should not be used in a combat situation where failure to chamber a round could have fatal consequences.

The no-go gauge measures the maximum headspace of the chamber. The bolt *should not* close on this gauge. If the bolt does lock, then the chamber is too deep or the bolt is undersized. Generally, it is best to avoid purchasing a barrel which accepts the no-go gauge. (You might be able to correct the problem with an oversized bolt, though the chances of coming across such a bolt are pretty remote unless you have a large number to try out in your barrel.)

It is also possible that a barrel which chambers the no-go gauge is set at the maximum chamber length. This type of barrel could be used, but maximum loads should not be fired in it since they might quickly create a dangerous condition of excessive headspace in the chamber. Brass fired from a rifle with the maximum headspace will stretch quite a bit, greatly shortening the reloading life of the empties.

To determine whether a rifle with excessive headspace (one that accepted the no-go gauge) is safe, use the field gauge. If the bolt turns on this gauge, *do not* ever fire a round in this barrel. To do so would be extremely dangerous.

If the chamber allowed the bolt to close on the no-go gauge but not the field gauge, it would be safe to fire the rifle with regular loads. You'd be wise, however, to check the chamber with the field gauge every few hundred rounds to be sure it has not enlarged sufficiently to be dangerous.

Whenever possible, buy only a rifle or barrel that accepts the go and field gauge but not the no-go gauge.

If you do a lot of reloading, the gauges can also be useful to make sure the chamber is not becoming enlarged through the battering caused by too heavy chamber pressures.

Even if you are on a limited budget, purchasing a set of gauges is an expense that can be a worthwhile investment. If you can only buy one or two gauges, consider first buying the field gauge so that you can avoid using a dangerous barrel. Then buy the no-go gauge so that you can discover when a barrel may be developing excess headspace.

If you change a bolt or barrel from one rifle to another, be sure to check the headspace with your gauges. The changeover might create a dangerous chamber space.

BARRELS

When purchasing a barrel, in addition to checking the headspace, you should look down the barrel toward a bright light to determine whether there are any "dings," craters, rust, or other flaws that don't belong in a good barrel. If the inside of the barrel looks bad, try cleaning it if possible since it may only be full of dirt. When you've determined that the inside of the barrel is free of flaws, take note of the barrel grooves: They should be fairly sharp rather than smooth and extremely rounded. If you should chance to see a "ballooning" on the inside of the barrel, avoid it since it has probably been fired with a bullet or dirt stuck in the barrel. Such a barrel is almost worthless—and probably unsafe.

If the inside looks good, check the outside. Remember that the grips will cover much of the barrel; blemishes on the barrel can be used to dicker on the price but should not be a cause for concern unless they are extremely bad.

Finally, sight down the barrel from several angles to be sure it is not bent.

Though it is hard to tell by just looking at a rifle, you should also inquire to determine whether the chamber and barrel are chrome-lined and what the twist rate of the barrel is.

If properly done, a chrome lining can facilitate cleaning the barrel. It may even head off some rust problems if you ever have to use the rifle under adverse conditions or fire ammunition with corrosive primers or powder. (This is not recommended, but may be necessary in battle or survival situations.)

Rounds are less apt to get jammed in a chromed chamber when the chamber gets dirty due to the use of poorly burning ammunition or inadequate cleaning of the rifle. (This is not to say that you should use poor ammo or not keep your rifle clean.)

Twist rates should also be taken into account when you buy a barrel.

A twist rate of "1 to 14" means that the bullet will make one turn for every 14 inches of barrel. This was the original twist rate and would indicate that you have a pretty old barrel. This does not pose a problem provided the barrel is in good shape and you will not be using bullets bigger than 55 grains (the current commercial size) or shooting in cold weather. When the temperature approaches the freezing point, a 1-in-14 twist does not create enough spin to keep the bullet accurate. Heavier bullets also suffer from lack of accuracy with this twist.

The 1-in-14 twist is the most deadly; at close ranges or in warm climates, you might wish to use a 1-in-14 twist.

The 1 turn to 12 inches is probably the best compromise for combat if you are using 55-grain bullets. It keeps 55-grain bullets fairly accurate while still allowing them to have enough yaw to "tumble" on impact and dump their energy with deadly results. The 1-in-12 twist can also be fairly good with heavier bullets in warm climates. In cold weather, accuracy will not suffer much except at ranges greater than one hundred yards.

The 1-in-9 twist is best with heavier bullets in the 60- to 70-grain range and will produce accurate shots with smaller bullets. The catch is that the bullets are extremely stable. Using soft- or hollow-point bullets would overcome this problem.

The 1-in-7 twist barrel is useful only if you will be shooting a lot of the new, heavier tracer ammunition. It is probably best avoided by most shooters.

When purchasing a barrel with a front sight base mounted (and this is the best route to go, since mounting the base can be a hassle), check to be sure that the base is not canted to one side or the other. The front sight should line up with the pin over the top of the chamber. It is easy to see whether the front sight base is in the right position by pushing the barrel into an upper receiver. When you sight down the rear sight area (even if the rear sight is not there), you can quickly tell whether the front sight base is canted. If it is, return the barrel to the seller for another (or do not buy it in the first place).

Several 16-inch barrel styles are available if you are building a carbine version of the AR-15. Probably the minimum useful length is eight inches. This produces a lot of flash and noise, as well as sending the bullet out at a velocity much less than that of a regular barrel. It is a handy length, but it requires a special permit from BATF since it is below the legal length of sixteen inches for a rifle barrel.

The next step up is the commonly available 10-inch barrel. To get around the BATF restrictions, a flash suppressor is often welded to the barrel to give it a legal overall length of sixteen inches.

A 16-inch barrel is available and probably the best choice if you are restricted by the legal-length specifications. Though the barrel is a little longer (a flash suppressor must be put on the end of it, adding an additional inch or so), it gives a higher velocity to the bullet and offers better accuracy. At the same time, the short, handy length of the carbine barrel is still maintained. A very good 16-inch, chrome-lined Colt barrel is available from Numrich Arms (West Hurley, NY 12491, 914/679-2417) for $135.

All short carbine barrels need a special short handguard and gas tube so be sure to order these if you are building a carbine version of the AR-15.

Heavy barrels offer a little more accuracy, especially with a bipod, but also create a heavier rifle. With the extra-long, 24-inch heavier barrel, several pounds are added to the rifle which ruins the "lightweight" handling normally associated with the AR-15.

Stainless steel barrels are best avoided unless you are going to be using the rifle in a really caustic environment. Though the barrels resist corrosion, stainless steel also holds heat longer than regular barrel steel. Shooting that does not create problems with a regular rifle can cause excessive barrel heat or even cookoffs with a stainless steel barrel.

FLASH SUPPRESSORS

Some sort of flash suppressor should be mounted on the rifle to protect the muzzle. Government flash suppressors are cheaper than others and work well, but there are some things they do not do well since they are a compromise version of a suppressor.

Flash suppressors are available which compensate for the upward/sideward motion of the barrel. These should get your consideration since they can make shooting a lot easier if you have to get off a lot of shots in a hurry. (See the accessories section for more information on flash suppressors/muzzle compensators.)

Choate Machine and Tool (Box 218, Bald Knob, AR 72010, 501/724-3138) produces several styles of flash suppressors which are not available anywhere else. One is a "slip over" hider that makes a 16-inch barrel look like a shorter barrel with a long carbine-style flash suppressor on it. (Though it is not actually useful, it does give you the "look" of the Vietnam-era carbines if that's something you are interested in.) This retails for $20.

The flip side of this is a flash suppressor that adds 5-1/2 inches to a barrel with a carbine-style flash hider. This allows you to make a "legal" 16-

inch barrel from an 11-inch one (provided you weld it on). This flash suppressor also sells for $20.

The maximum reduction of flash for nighttime firing can be achieved by using the extra-long, fluted "M14" style flash suppressor from Choate. It, too, costs $20.

INSTRUCTIONS AND TIPS

When you buy a bolt carrier, you should check it to be sure it is chromed where the bolt rides it. This, too, will aid in preventing jams.

All-chrome bolt carriers are sometimes available, but the chrome is not needed anywhere except the bolt area. Besides being unneccssarily expensive, the extra chrome actually causes extra wear on the inside of the upper receiver and may shorten its life considerably.

If you are building a rifle with a forward assist, you will need a military M16-style bolt carrier with forward assist notches cut along the side with the two oil holes (ejection port side).

A rifle which will operate as a selective-fire weapon will need a bolt carrier that has the same length of metal on top and bottom as measured from the rear of the bolt carrier to the cut-out area. Bolt carriers designed for the commercial "Sporter" AR-15 will not work for selective-fire use; they will have a much shorter stretch of metal on the lower rear of the carrier than on the top.

Chromed bolts are probably best avoided. For some reason, the chrome seems to flake off the bolt, changing its outward dimensions as well as making a mess in the chamber with heavy use. Additionally, they look terrible when they are half chromed and flaky.

Chromed firing pins are about the only type on the market, but you should avoid nonchromed firing pins.

Stainless steel or steel upper/lower receivers are sometimes seen on the market. These are expensive and make the rifle heavier while offering little in the way of advantage over the aluminum receivers. Though it might seem that a steel receiver would create a safer rifle, this is not the case. The bolt locks into the barrel extension. The receivers need not be excessively strong to be reliable. (In fact, the receivers could probably be made of some of the new space-age plastics. It is doubtful that this will happen in the near future, however, since shooters would be a little leery of such a product.) Currently, the non-Colt receivers seem to be of very high quality.

Though there are some other specialized tools for working on the AR-15, the only other ones you'll probably need are regular shop tools: a vise, vise-grip pliers, wire cutters, needle-nosed pliers, regular pliers, files, a wood rasp, drill bits (without the power drill), a grinder, a hacksaw, a small hammer, and a set of files. Drift punches are handy to have, but you can get by with a nail set and/or several small nails. File the points off flat on the nails and you will have a set of inexpensive drift punches.

An ice cube tray or some small plastic containers are a must to keep track of some of the smaller AR-15 parts as you assemble the rifle. If you just scatter parts over the tabletop on which you're working, sooner or later you'll be groping around on the floor looking for some small, essential part that has rolled off the table.

If you are not hampered by eyes that wear tempered glasses, you would do well to get some safety glasses and use them. Even though you might think you will never need them, wear them anyway.

The fitting work needed to get things together varies from a little to none at all. The areas that are often the worse are the small holes in the receiver. Since some manufacturers use a baked-on finish on the receivers, the thickness may vary somewhat according to the depth of the finish. This is of little significance *except* in small holes on the receiver where the finish may make the dimensions small enough to trap a spring or part.

When you assemble the rifle, always be cautious about pushing a part into some of the small, "dead-end" holes on the lower receiver. If the fit seems tight, the holes may need to be reamed out.

A file can be useful for reaming out the holes (often the handle end is better than the teeth end). The soft metal of the receiver is very easy to remove, so do not be overly conscientious when reaming.

Drill bits can be used to remove small amounts of metal from a hole. Find a bit that is a shade larger than the hole, and hold the bit in a pair of pliers. Rotate the bit in the hole by hand, and try the part in the hole from time to time to be sure you are not exerting too much pressure. Do not use a power tool.

When the hole is big enough, be sure to empty out the little bits of metal filings so that they will not bind things up.

Roll pins are often a headache. They need to fit

tightly so that the rifle does not shake itself apart during recoil. It often seems that roll pins are about two sizes too big. To get the pins in, file the end of the pin into a rounded point, and maybe even squeeze it together with some pliers. If the pin still refuses to start in its hole, then the end of the hole may be reamed out just enough to get the pin started.

Though gunsmithing books would lead you to believe that you just rap on a roll pin once or twice and it's in, that is not the case with most AR-15s.

Generally, a hammer is called for and usually a drift punch (the type with a small nipple in its center) is the best. Be sure not to overdo things since the receiver is soft (the barrel and bolt take all the pressure during firing). Though I have never heard of a piece breaking off, you should use a series of gentle taps rather than one or two crunches. This is especially true in areas like the bolt release and the trigger guard where the metal is thin and unsupported.

When you are putting in the pins that hold the trigger or hammer in place or any of the roll pins, here is a trick that is often a real time-saver: Find a push punch, nail, or even a part of the rifle that freely fits into the hole that the pin will be going into. Position the piece in place, and use the punch (or whatever) to hold it in place. Now push the pin in from the opposite side. When the pin goes on through, it will push the punch out of place.

When you are using pliers, hammer, files, or other tools that may take a chunk off the finish of your rifle, try to be careful so that the rifle does not get damaged. Masking tape or Scotch tape can be put over the surface of the rifle in areas where the tool may mar the finish on the rifle. A small piece of leather or a plastic bag can also be very useful in protecting parts from being marred by tools. It is better to take a little time to protect the rifle than to try to touch it up later.

The aluminum in the receiver cannot be easily touched up if it gets dinged, but flat epoxy paint or enamel paint can make things look a lot better. Cover the spots where the aluminum shows through the finish. It is next to impossible to cover a large area with paint and still make it look good.

For large areas on the receiver which need to be refinished, use Gun Kote. This will produce a good-looking finish, and it can be applied to the barrel or other areas as well as the receiver.

Gun Kote is a phenolic resin base with molybdenum disulfide (a lubricant) suspended in it. The Gun Kote is sprayed (from an aerosol can) onto the metal that has been heated to 180 degrees Fahrenheit and then baked in an oven for half an hour at 300 degrees. This creates a tough coat on the metal that is similar in appearance to parkerized metal. (A material identical to Gun Kote is often used by the Navy and Marines for refinishing their M16 rifles.) The real plus is that it can be done in your kitchen without the need for special equipment, and there are no dangerous chemicals.

Since you will be heating up the rifle to a point that might melt some plastic parts, be sure to remove the stock, pistol grip, handguards, buffer, bolt (it may have a nylon part inside the extractor spring), or other parts which might be damaged by heat.

Gun Kote is available from M.A., Inc. at $5.50 for a 12-ounce aerosol can (enough for completely refinishing three AR-15 rifles).

At the time of this writing, Buckeye Chemical Products (P.O. Box 7869, Oregon, OH 43616) is working on a refinishing product for aluminum which should soon be available. You might wish to drop them a line to see whether it is yet available.

Steel parts (everything except the receivers and plastic furniture on the rifle) can be darkened with touch-up gun blue. (This chemical is also called cold blue since no heat is needed to use it.) Though the color (a dark blue-black) may not match as well as the original, it is generally close enough to hide any scratches or nicks from all but the minutest inspection.

Cold blue makes it easy to blue a part; it can be purchased in most gun stores. Touch-up blue is especially useful to make the roll pins look right after you have gotten them into place.

After you use touch-up blue, be sure to carefully wipe it off the part and oil it. The "bluing" is similar to rust, and rust often "grows" in or around an area several hours after a part has been blued. Check it again several days after it has been blued. If there is any rust, lightly burnish it with fine steel wool and cover the area with a thin coat of oil.

Loctite is a slow-setting glue that can be very useful for making sure that parts won't come apart during the hammering of recoil. Be sure you use only a small drop at a time, allowing twenty-four hours for the material to harden before you put stress on any part. Also be sure you never disassemble the parts joined by it: When Loctite is in a threaded area, it will permanently keep the parts together. Though the glue can be dissolved with acetone (nail polish remover), getting the acetone to seep into a tight area is often quite a trick.

Though Loctite works well, Gun-Tite is a special type of Loctite formulated just for guns. You may be able to find it at your local gun dealer.

Care and patience are needed in assembling the rifle. Do not try to rush, and everything should go together. The following procedure list will help you. You may find it helpful to check off each step as you go along. Do not skip around as some steps must be completed before others if the rifle is to go together properly. Check the diagrams to be sure you have the right part and have it oriented properly. Even better, borrow a friend's AR-15 if you can do so.

10. Assembly

The AR-15 is just one step away from being an automatic weapon. The one essential part needed to make it function reliably in the automatic mode is the sear.

When you order a parts kit, the military automatic parts are currently the most common in the marketplace and are probably what you will receive. These parts work well. The automatic sear is usually missing from the kit so that you should not run into any problems with the BATF regarding possession of an illegal automatic weapon.

One small modification to the disconnector must be made, however. Without modification, the rile could be forced to fire in an automatic mode if you have the military-style automatic hammer, trigger, selector, and disconnector.

This can happen if you force the selector into the automatic position and are using ammunition that has sensitive primers. After the first shot, the disconnector will not catch the hammer. If you continue to hold the trigger, the hammer follows the bolt forward, slamming into the firing pin. The pin in turn hits the primer of the cartridge as it is chambered.

Though the bolt of the AR-15 is designed to prevent the ignition of a primer before the bolt is locked into position, the force of the hammer riding behind the firing pin is enough to at least partially fire a primer. You might then get a string of shots, but the bullets will not be traveling out of the barrel at their normal velocity because the primer won't be setting off the full charge of powder.

The end result will be that sooner or later a bullet will not exit the barrel, and the next round to fire will blow up the barrel!

Therefore, you not only will have an automatic weapon that is illegal as far as the BATF is concerned, but you will also have a rifle which is ill-suited for shooting. The BATF does not have to prove the gun is suitable or safe to use in the auto mode; all they have to prove is that it *can* fire more than two rounds at a time with one pull of the trigger. They can also use special rounds with extra-sensitive pistol primers or the like.

STEP-BY-STEP INSTRUCTIONS

You can avoid future problems with the BATF by grinding off the tail of the disconnector. A file won't work for this procedure, since the metal of the disconnector is nearly as hard as the file. Use a grinder if one is available. In a pinch, you might get by with a whetstone and/or hacksaw.

The receiver's tail (located behind the notch over the spring hole) must be removed. This procedure will keep you legal and safe. Do this as soon as you get the part so that authorities will know you had no crime in mind before you assembled the rifle. If you use a grinder, be sure not to overheat the disconnector since it will ruin the temper of the metal. As you work, let the piece cool off occasionally, or dunk it into cold water.

Other steps needed to assemble an AR-15 are as follows (refer to Diagram 2):

1. The magazine catch (2-16) is pushed through its well in the lower receiver while the catch spring (2-17) is held in its well by the catch button (2-18). Once the threads are engaged, turn the catch until it is threaded into the button. It will be necessary to push the button into its well below the surface of the receiver with a tool. The base of the catch should be even with the button's face when you're done. Depress the button and look down the magazine well to be sure the catch will not block the well when the button is depressed.

If the catch does block the well when the button is depressed, unscrew the latch a turn until it stops blocking the magazine well when the button is depressed. The catch should not bind in the receiver. If it does, it may be necessary to remove it and smooth off any burrs or high points that are causing it to get hung up.

2. Install the bolt catch spring (2-15) in its hole along with the bolt catch plunger (2-14).

3. Set the bolt catch (2-13) into its spot and drive the roll pin (2-12) home. (The general practice is to drive roll pins into their holes so that they are slightly lower than flush with the opening of their hole.)

4. Pop the trigger spring (2-10) on the trigger (2-9). The spring has its crosspiece to the front and under the front bar of the trigger. Each loop goes around the outside projection hub on either side of the trigger.

5. Place the hammer spring (2-4) on the hammer (2-3). The two legs of the spring point toward the base of the hammer, and the crosspiece is behind the neck of the hammer. Check the diagram for proper orientation.

6. Set the disconnector spring (2-10) in its well in the trigger. The small end of the spring points down toward the trigger.

7. The disconnector (2-8) is now positioned on its spring in the slot of the trigger. To speed up the assembly, temporarily slip the selector lever detent (2-23) into the pin hole to hold the disconnector and trigger together. The thickest part of the detent should be positioned so that it is inside the center of the disconnector's hole.

8. With the detent holding the disconnector and trigger together, lower them into the lower receiver so that the trigger spur goes through its slot in the base of the receiver.

9. Get the pivot hole of the trigger lined up with the hole in the side of the receiver. Drive the trigger pin (2-2—it's identical to the hammer pin) in from the side. Gently tap the pin (so the detent does not get damaged) until it is driven almost all the way in. The detent will be driven out by the pin (be sure not to let the detent get away). Just before you drive the pin on through the receiver, line up the trigger with the receiver and tap the pin on through.

10. Push the hammer into place. The legs of its spring go over each hub of the trigger. Hold the hammer down and insert a nail or punch through the receiver hole. While the nail holds the hammer

and its spring, drive the pin into its hole from the opposite side of the receiver.

11. Cock the hammer back so that it is held by the trigger in its cocked position.

12. Wiggle the selector (2-7) into its hole.

13. Check the dimensions of the buffer retainer (2-38) in its hole. If the hole is too small, ream it out as described above with a file or hand-held drill bit.

14. Check the buffer tube (2-37) by screwing it into the receiver. If the buffer tube has two holes in its side, push a screwdriver through the holes and use it as a lever to screw the tube into place. If the buffer does not have side holes, it will probably have a square end that will allow you to use the armorer's wrench or crescent wrench to screw it into the receiver. When it is screwed into the rear of the receiver, it should just cover the edge of the buffer retainer hole. If it goes too far into the hole, some of the threaded edge may have to be carefully removed.

15. On the shorty telescoping stock, be sure the locking nut is adjusted so that the stock can be positioned correctly; be sure, too, that the end plate for the receiver is in position when you do this. If you want the stock to be the minimum length, carefully remove some of the threaded end and file the threads back into the end of it where the cut was made. Generally, it is easier to leave the extra length on the telescoping stock; most shooters find it more comfortable to use with the extra 3/4 to 1 inch of stock left on.

16. When the dimensions are correct, insert the buffer retainer spring (2-39) and the retainer into their hole.

17. With the telescoping stock, the rear push pin detent (2-34) and its spring (2-33) are located in their holes at this time after the dimensions of the detent hole have been checked to be sure it is big enough.

18. With the telescoping stock, be sure the receiver plate is in place. As the telescoping stock is screwed into the receiver, be careful not to pinch the detent spring. Be sure the buffer retainer is held under the front edge of the buffer tube.

19. With the regular buffer tube, be sure the retainer is held down as the tube is screwed home.

20. If the buffer tube holds the retainer in its hole while still allowing its narrow nub to stick up, back the buffer up a half turn or so and apply one *small* drop of Gun-tite or Loctite to the threads of the buffer tube. Tighten it back to

where it belongs. Be careful not to get Gun-tite onto the retainer. (Military armorers often drill a hole into the buffer tube/lower receiver and pin the buffer tube into place. Loctite is a lot easier.)

21. With the regular stock, set the rear push pin detent (2-34) and its spring (2-33) into their hole *after* checking the dimension of the hole to be sure the detent can move in it freely.

22. With the regular stock, push the stock (2-32) onto the buffer tube while being careful not to pinch or bend the end of the detent spring sticking out of the back of the receiver.

23. With the regular stock the trap door assembly can now be screwed onto the stock along with the lower swivel. (Many users may prefer to put the swivel on backwards so that it is less apt to catch on equipment or jacket "slash" pockets.)

24. If your stock is one of the old-style stocks without a trap door, assemble the rear swivel (2-31) to its post (2-32) with its pin (2-30). Mount the unit in the stock with its pin (2-28), and fasten the stock to the buffer tube with the stock screw. Be careful not to pinch or bend the end of the detent spring sticking out of the back of the lower receiver.

25. Push the buffer spring (2-24) into the buffer tube. Bending it upward as you snake it into the tube allows it to clear the buffer retainer.

26. Slip the buffer (2-25) into the buffer tube. The nylon tip goes toward the rear of the stock and the large, flat end toward the barrel end of the stock. You may be able to wiggle the buffer slightly to get it past the buffer retainer, or it may be necessary to depress the retainer. Once the buffer is in the tube, the retainer should hold it in place.

27. Orient the trigger guard (2-41) into its position on the magazine well, and depress the spring-loaded pin on the guard's front. This will allow you to slip the guard into its niche behind the magazine well.

28. Swivel the rear of the trigger guard into position and drive the pin (2-40) home to secure it.

29. With a small screwdriver or similar tool, reach through the rear push pin hole from the safety selector side of the rifle and push the rear push pin detent back into its well. While holding the detent back, wiggle the rear push pin (2-35) into its hole. Be sure to line up the slot on the pin with the detent.

30. Using the same technique, set the front detent spring (2-33) into its well and then the detent (2-34) into place. Hold them down with a small tool pushed through the pin hole. Push the front push pin (2-1A) into its spot.

31. Slide the pistol grip onto the lower receiver to be sure the grip fits. If it does not, it may be necessary to remove plastic from the grip with a file or rasp. Once it fits, set it aside for a few moments.

32. Make sure the selector is all the way into the lower receiver, and turn the receiver upside down. Drop the selector detent (2-23) into its well (pointed end toward the selector) and slide the detent spring (2-23) into the well behind it.

33. Set the grip lock washer (2-20) on the grip screw (2-19) and drop them into the pistol grip (2-21) so that the threads extend through the hole.

34. Slide the grip into position while being sure that the selector detent spring goes into its hole in the grip. Extreme care must be taken not to bend the spring between the receiver and grip. When the grip is on the receiver, tighten the grip screw. On some grips, a small space may be left between the rear of the grip and the receiver. Tightening the grip screw an extra bit will help get rid of this space, but be careful not to use excessive pressure since the grip can break or the screw may strip the threads of the aluminum receiver.

35. If the selector works too hard, it may be necessary to remove some of the pistol grip and shorten the selector detent spring about a turn. Use the above procedure to remount the grip.

36. If the selector rubs against the receiver when the selector is moved, it may be necessary to grind off part of the receiver or the lower edge of the selector. (This will look better since shiny metal exposed on the selector can be darkened with touch-up blue.)

37. If you are assembling an automatic version of the AR-15, set the auto sear (2-6) into position and secure it with its pin (2-5).

The lower receiver group is now assembled along with the stock.

38. If the bolt carrier (1-13-D or 1-13-E) does not have the key (1-13-C) mounted on it, position it on the bolt and screw in the two hex bolts (1-13-B). A few drops of Gun-tite can be used on the threads of the bolt. If you do not use the Gun-tite, use a nail set to peen the bolt carrier metal over the edge of each hex bolt.

39. If the bolt (1-4) is not assembled, snap the

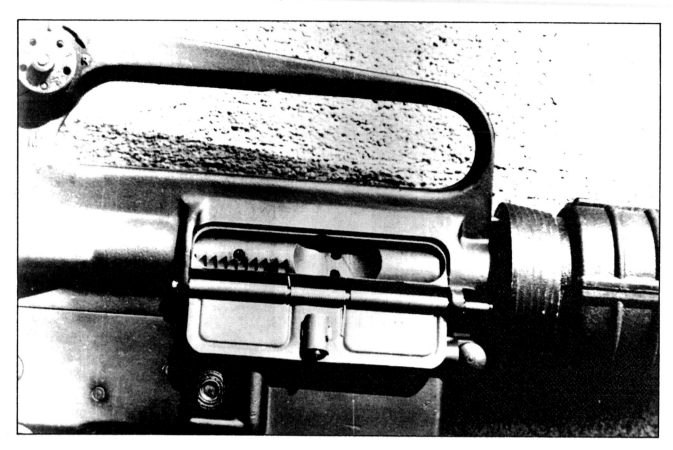

Proper assembly of ejection port cover

three bolt rings (1-11) onto it. These rings *must not* be lined up with their spaces in a row since to do so will allow gas to seep through the space and hinder the action of the rifle.

40. Set the extractor spring (1-7) in its well. Many kits have a small silicon insert that goes inside the spring to aid in its functioning.

41. Put the extractor (1-6) in its proper location and slide its pin (1-5) into its hole.

42. Notice that the ejector (1-9) has a slot in it. The longer end above the slot goes toward the face of the bolt, and the slot lines up so the roll pin (1-8) can go through the slot. Put the ejector spring (1-9) into its well and the ejector after it with it properly aligned (an empty cartridge can be used to hold the ejector down). Push the drift punch into the roll pin hole to retain the ejector. Drive the roll pin into the bolt from the side opposite the punch.

43. Be sure the extractor is aligned so that it is on the same side as the bolt carrier's two oil holes (and bolt assist grooves), and push the bolt into the bolt carrier. (If you don't align the bolt properly, it is impossible to push the cam into place in the next step.)

44. Push the bolt cam (1-3) into the bolt and bolt carrier. The cam should then be turned so that it is under the bolt key.

45. Slide the firing pin (1-2) into the bolt by inserting it into the rear of the bolt carrier and bolt.

46. The firing pin retaining pin is now slipped into the bolt carrier from the side opposite the bolt's extractor. Do not spread the feet of the pin. If the bolt carrier is turned with the bolt faceup, the firing pin should not fall out. If it does, remove the retaining pin and reinsert the firing pin.

Not all upper receivers have a forward assist. Therefore, you may need to skip the next few steps if your rifle does not have this feature.

47. The forward assist assembly (3-32 through 3-38) is generally together when purchased. If not, note the alignment of parts on the diagram and assemble them accordingly.

48. Push the plunger spring (3-33) around the forward assist and shove the forward assist assembly into its well. If the forward assist is the round style, be sure the slot on the assembly is positioned so that the retaining pin (3-32) can go by it. Secure the assembly with a drift punch.

49. Drive the retaining roll pin (3-32) into its hole from the lower side of the receiver.

50. Set the rear sight leaf spring (3-31) into its well in the receiver. The "bump" of the spring should be in the center with its two ends pointing downward.

51. The rear sight (3-30) should be held down on the rear sight spring, and the windage screw (3-29) pushed through the windage screw and receiver holes. When the threads of the windage screw are up to the side of the receiver, screw it into the receiver. Stop screwing when the sight is about halfway between the receiver ears.

52. Drop the detent spring (3-28) and detent (3-27) into their well and push the windage knob (3-26) over them. Hold them in position with a small punch or wire.

53. Drive the roll pin (3-25) into the windage knob.

Getting the ejection port cover and its spring into the proper orientation is very awkward and will probably require several tries. The assistance of a "helper" can greatly simplify matters at this point.

54. Snap the C-spring (3-22) onto the ejection port cover pin (3-21) if it is not already on the pin.

55. Run the cover pin into the receiver from the barrel end of the receiver and into the right half of the ejection port cover (3-24).

56. Position the spring (3-23) so that the tail on the barrel end of the cover is on the inside of the cover (the side with the latch on it).

57. Using a pair of needle-nosed pliers, twist the spring an extra turn so that it is tighter. It should end with its tail on the stock end of the spring against the receiver. Hold it in this location and slide the pin on through the cover and the hole on the opposite side of the port.

58. Carefully lay the upper receiver aside until the barrel is ready to be slid into it. Remember that the ejection port cover is not locked onto the receiver until the barrel is mounted. Do not let it slide out.

The front sight (3-5) will usually be mounted on the barrel along with barrel nut (3-19), handguard cap (3-14), sight pins (3-6), and possibly the front sight post (3-11), its spring and detent, and the front swivel. If these parts are on the barrel, some of the following steps may be omitted.

59. Slide the barrel nut (3-19) onto the barrel from the muzzle end of the barrel. Be sure the teeth end of the nut is toward the muzzle.

60. Slide the handguard cap (3-14) onto the barrel from the muzzle end.

61. Slide the front sight (3-7) onto the barrel from the muzzle end.

62. Tap the front sight pins (3-6) into the sight base from the right side of the sight. The tapered end of the pins goes in first.

63. Set the front swivel (3-10) into the proper location and drive its roll pin (3-9) home. If the swivel is held in place by a rivet, you'll need to peen the rivet closed.

64. Drop the front sight spring (3-13) into its well.

65. Push the front sight detent (3-12) into the well and screw the front sight post (3-11) into position.

66. After putting the flash suppressor lock ring (3-4) onto the muzzle, screw the flash suppressor (3-3) onto the muzzle.

67. Slide the slip ring (3-20) over the end of the barrel. It should have room for the weld ring (3-18). If it does not, it is on backwards.

68. Insert the weld ring (3-20) into the slip ring.

69. Use needle-nosed pliers to pull the C-ring (3-17) onto the barrel nut. Start on one side of the spring and work your way around the ring. Be sure the spring is in the slot of the barrel nut.

70. Wiggle the barrel into the receiver. If the "bump" of the barrel does not go into the slot of the receiver, some material may have to be removed from the receiver. Don't remove too much; a tight fit is desirable.

71. If the barrel is too loose in the slot, use a small wire to make a shim that will fill in the space so that the barrel will be tight when it is pushed onto the receiver. There should be no play in the barrel when it is twisted after the "bump" is in the receiver slot.

72. Screw the barrel nut onto the receiver. An M16 wrench is almost essential for this step, and an aluminum barrel clamp in a vise is also very helpful. The nut should be tight, but not extremely tight. (If you have a torque wrench, the barrel nut is normally 35-45 foot/pounds.)

73. Use a small tool to align the slip ring, C-ring, and weld spring slots so that they are even with the gas tube hole on the front of the receiver.

74. Back up or tighten the nut so that the receiver hole is lined up with one of the notches of the nut. A small tool may be useful to check the alignment after you slip the bolt carrier and bolt

into the receiver. The tool should line up with the carrier key, C-spring, weld spring, slip ring, and one of the cutouts in the barrel nut. When it is aligned, remove the bolt carrier.

75. Push the gas tube (3-16) through the nut, slip ring, weld spring, and C-spring into the receiver. Continue to push it into the receiver until the end of it clears the front sight base.

76. Check the position of the holes of the gas tube so that the hole going through it is on either side of the barrel and the gas hole points toward the barrel. Pull the gas tube back out of the receiver part way and into the front sight base. Be careful not to bend the tube.

77. When the holes of the tube are lined up with the holes in the front sight, drive the roll pin (3-15) through the sight base and tube.

78. Push the bolt carrier and bolt into the upper receiver. The bolt carrier key should slide over the gas tube with very little resistance or friction. If the key does not go around the tube without binding, check the alignment of the tube. A screwdriver can be used to align the end of the tube toward the center of the receiver. Also check the key; polishing its inside slightly with a file may allow it to go over the tube easily. Do not try to fire the rifle if the key does not fit the tube loosely. After the tube is aligned, remove the carrier.

79. Stick the front of one of the handguard halves (3-1 or 3-2) into the front handguard cap. Keep in mind the fact that with many AR-15s there are right/left or upper/lower halves which cannot be substituted for each other.

80. Push the slip ring back toward the receiver and then pull the handguard half into position. It may be necessary to use a screwdriver to lever the slip ring back from the handguard. If the handguard is slightly too large to fit, use a file or wood rasp to remove a little of the outside edge of the areas which seem too large.

81. Repeat the above two steps to get the other handguard half into position.

82. Pull the two push pins (2-35 and 2-1) out. Put the upper receiver into position and shove the front push pin through the two receivers.

83. Push the upper receiver down and push in the rear push pin if possible. If the rear push pin does not go through the rear receiver easily, remove metal from the rear push pin hole of the *upper* receiver. A round file is most useful for this task. Be careful in so doing, since it is impossible to put the metal back!

84. When the rear push pin fits according to your taste, leave the halves open and slide the charging handle (1-14) into the upper receiver. To do so, push it into the large tubular hole and move the lugs on the inside of the receiver up through the two cut-out grooves in the receiver slot. Don't push the charging handle all the way in.

85. Insert the bolt carrier into the upper receiver, and shove it and the charging handle all the way in.

86. Close the receivers together and push in the rear push pin.

87. Insert an empty magazine into the magazine well and be sure it is locked into place.

88. Pull back the charging handle and release it. The bolt should be held back and the charging handle should not be carried forward unless the bolt release on the side of the receiver is pushed. If the action closes rather than staying open, pull the charging handle all the way back again and look into the ejection port while the action is being held back. The bolt should be behind the bolt catch. If it is not, then the buffer spring needs to be shortened.

89. If the buffer spring needs to be shortened, hold the buffer in its location while you depress the retainer (2-38). When the buffer is released, carefully remove it and its spring. Cut off one-quarter turn of the spring and replace it and the buffer, and repeat the above step. Do not take off too much of the spring without testing it for the proper length. (The spring won't grow back if you cut off too much.)

90. Try cycling some dummy rounds through the rifle to be sure it functions properly.

91. If you have headspace gauges, check the chamber before firing the rifle. (Check the methods of using the gauges as outlined elsewhere in this book if you are unsure of their proper use.)

92. Lightly lubricate the bolt and all moving parts. (Study the lubricating recommendations elsewhere in this book.)

TEST FIRING

The next step is to test-fire the rifle. Care should be taken to avoid the use of maximum loads until you have test-fired your rifle a number of times. Start out by placing a single round in the magazine and cycle it into the chamber. Keep the rifle pointed in a safe direction since it might fire the round if the firing pin should be too long.

Examine the empty brass after the first firing. Is

it ruptured? Does the head (primer) end appear deformed?

If all seems all right, try firing groups of two or three rounds. Again, examine the brass to be sure that it is not showing problems that may be developing with the rifle.

If there seem to be any problems, do not hesitate to take your rifle to a competent gunsmith. He can make your rifle shootable for a very small price. A small sum of money spent to make your rifle safe and usable is certainly preferable to getting injured because the rifle you assembled does not work quite right.

If you take care and follow each step without hurrying, you should be able to assemble a quality weapon that will rival the best you can buy.

To finish up on the rifle, use touch-up blue (for steel parts). The blue will not be effective when used on aluminum parts, such as receivers and a few of the lower receiver parts. You can use black paint to give your rifle a brand-new look, but the best bet is to use Gun Kote.

POSSIBLE DIFFICULTIES AND CHANGES

A problem which may develop over time is "walking pins." When the rifle is fired a number of times, the recoil and action movement cause some hammer and/or trigger pins to slowly come out of one side of the lower receiver. Though the rifle will generally continue to function and the pins rarely come out completely, it is not reassuring to look at the rifle and see an open hole on one side and a pin half way out on the other.

The cheapest solution is to remove the pin, place it in a vise, and lightly peen the end of the pin so that it is larger than it would normally be. Place it back in the receiver from the side from which the pin was pulling out (where it was leaving an opening). Push the regular end of the pin in first and then drive the peened end into the receiver.

Another solution is to turn new, oversize pins on a metal lathe. This entails a lot of work, and the receiver holes will become oversized after the turned pins are installed, thereby making the use of regular-sized pins subsequently impossible.

The easiest solution is to purchase "Anti-Walk" pins. These pins have an oversized head that keeps one end of the pin from going through the receiver. After the pin is pushed through, a C-spring goes around the other end of the pin to lock it in place. (This is similar to the arrangement on the AR-180.)

A set of two Anti-Walk pins are available for

$12.50 from L. L. Baston Company (Box 1995, El Dorado, AR 71730).

If you will be using your AR-15 for target work (most stock rifles are capable of three minute of angle, or m.o.a., or better with good ammunition) or wish to have additional accuracy when using the iron sights, there are several things you might consider doing *if* you are handy at metal work.

Many users find the standard front sight post too wide and are bothered by its taper from bottom to top. If you find yourself in this group, it is possible to place the sight in a small lathe (or drill press if you are really careful), and turn it down with a hand-held file so its diameter is around 0.05 inch and the taper gone.

However, if you do so, you will find the rear sight aperture too large! Your next step will then be to silver-solder a small piece of metal to cover the inside of the sight (the side away from you when you shoot), and redrill a hole of about 0.037 inch in its center.

If you prefer not to redrill the rear sight, you can also grind off the front sight post and silver-solder a new bar onto it. The bar is then turned down on a lathe with a hand-held file until it is square.

Do not tackle this job unless you know exactly what you are doing, since to make an error means you will be buying some new parts. If you change the rear sight, you will probably need to plan on using just one sight rather than flipping to the long-range sight, since the sights probably won't stay on zero when you change from one to the other after you have redrilled the peep hole. You can still shoot at ranges beyond your zero if you compensate by aiming over the target according to the target's range. (The term for this procedure is hold-off. You adjust for bullet drop by changing the point of aim rather than the sight setting.)

Another rather complex change that can be made to the sights is to change the front sight and windage drum so that they will give half-minute adjustments rather than full-minute ones.

To do this, an extra hole is cut between each hole in the drum and a new notch carefully filed into the front sight between each of the five existing notches.

As previously mentioned, these tasks are not to be undertaken by the novice in gunsmithing work. They are not needed unless you are trying to get as much accuracy from your AR-15 as is possible. In general, a better practice would be to go to a rifle scope if you feel more accuracy is needed.

11. Conversion Kits and Modifications

A number of "kits" or modifications are available on the market which allow users of AR-15 rifles to fire ammunition other than the 5.56mm/.223 Remington. It is therefore possible to fire a number of different rounds in the AR-15 rifle, including .22 LR, 9mm Luger, and .45 ACP. It is very possible that other pistol calibers may be on the market in the near future since the kits usually use blow-back actions which are quite simple and safe for most pistol rounds.

These adapter kits allow a lot of advantages to the shooter, similar to those of the "family of weapons" or systems concepts. You can use a caliber to suit your needs while not having to switch over to a system of cocking/safety/aiming of an unfamiliar firearm. The advantages of kits include:

- The lower recoil of lighter rounds the kits are chambered for is good for training new shooters.
- The pistol rounds are more suitable for some types of use for which the 5.56mm is overpowered, such as urban police work or indoor self-defense.
- No FFL is needed to purchase a kit, and the other rounds, especially the .22 LR, make practicing with the AR-15 a lot cheaper due to the reduced ammunition costs.
- You can own a number of rifles for the price of one or two complete rifles.

.22 LONG RIFLE CONVERSION KITS

The .22 LR conversion kits are perhaps the best buy since only the bolt/bolt carrier and magazines are replaced in the rifle; the barrel is of the correct caliber for the .22 bullet.

In the past, a number of groups have developed .22 LR conversion units: Colt Firearms, the U.S. Rock Island Small Arms Laboratory, Military Armaments, U.S. Armament Corporation, SACO (of the Maremont Corporation), and Bro-Caliber International. There are two .22 LR kits on the market at the time of this writing: the M261 military kit (available on the "surplus" market) and the Bro-Caliber International unit.

Like the other self-loading adapter units on the market, those for the .22 are simple blow-back actions (as are most semiauto .22 rifles) with a chamber insert that is shaped like the end of a 5.56mm empty case. This insert fills out the space in the chamber, while the unit's inside becomes the new chamber for the .22 rounds.

Both units often suffer from reliability problems. At their best, neither offers 100 percent reliability. From a training standpoint, this is perhaps not all bad since it forces the user to learn to recover quickly from problems which might occur under the worst of circumstances. If the rifle is being used for hunting small game or the like, functioning problems can be aggravating to say the least.

The Bro-Caliber unit seems to suffer from quality-control problems. Some units function well, while others practically need to have each round fed through them by hand. If you get a Bro-Caliber unit, test it out right away to be sure it works. Return it for another if you encounter any major problems with it.

The Bro-Caliber kit is available in two models. Kit 1 is for semiautomatic use, while Kit 2 allows the owner of an automatic version of the AR-15 to use the .22 LR unit in the auto mode.

The Bro-Caliber has its own magazines which look markedly different from the regular 5.56mm magazines of the AR-15; there is no mistaking the thin magazines for regular magazines. (This feature might make a difference to those who want the weapon to look like a regular AR-15 when it is in use with a .22 LR conversion.)

Bro-Caliber's semiauto kit comes with one 10-round magazine and costs $130; the selective-fire kit comes with a 30-round magazine and costs $142. Additional magazines are available for $18 for a 10-round magazine and $23 for a 30-round magazine. Kits and magazines can be ordered from Bro-Caliber International (1105 Hulman Building, Dayton, OH 45402).

The M261 unit fires only in the semiauto mode and uses magazines that are nestled inside stock AR-15 magazines (you furnish the AR-15 magazine for the .22 magazines to go in). Though the .22 magazine units can be placed in or out of the standard magazines so that the magazines can still be used with the 5.56mm rounds, it is a bit of a hassle. A better practice is to use defective 5.56mm magazines, put the .22 magazine inserts into the defective magazines, and leave the inserts there.

The M261 unit is currently available from SACO, Inc. (323 Union St., Stirling, NJ 07980, 201/647-3800) for $178. The kit comes with three 10-round magazine inserts and a few spare parts. Occasionally Numrich Arms Corporation (West Hurley, NY 12491, 914/679-2417) also has the kits for $175 for a new kit or only $130 for working but slightly blemished kits (a good savings since the finish on the M261 is mostly paint which quickly nicks off anyway).

Some users of the M261 have found that it can be made more reliable by keeping it religiously clean (especially the chamber) and by modifying the firing pin.

The modification of the firing pin consists of grinding the point down so that an "I" area (rather than an "0") hits the edge of the rimfire round. Care must be taken when changing the firing pin so that it does not puncture the brass (which might cause gas to escape back into the action), or make the firing pin miss the primer area in the rim of the round.

Another essential modification is to have the chamber "throated" (just like a combat automatic pistol) by a gunsmith. Doing so will end a lot of feeding failures with most of the .22 conversion kits.

Both .22 conversion kits should be carefully cleaned for best functioning.

SINGLE SHOT FIRING

Special inserts are also available which allow .22 CB Cap, .22 Short, .22 Long, .22 Long Rifle, or (with a second adapter) the .22 Magnum to be fired—single shot—from the AR-15. These inserts are shaped on the outside like an empty 5.56mm round (complete with the extractor groove), with a chamber for the smaller round to be fired in it cut out of the inside of the adapter. A small striker/insert is placed into the adapter behind the round; this striker transfers the firing pin's force to the rim of the round in the adapter.

The round to be fired is placed in the appropriate adapter, the action of the AR-15 is pulled open, the adapter with the round is placed in the chamber of the rifle, and the action carefully closed. (WARNING: *Do not* let the action slam shut on the round; it may be fired by the AR-15's floating firing pin.) The rifle is now ready to shoot the single small caliber round. Pulling the action open after firing extracts the adapter, and a small stick is used to push out the empty cartridge and striker from the adapter.

One of the nice features of the single-shot adapter is that it allows the shooter to go from 5.56mm to a smaller caliber and back without opening the receiver and pulling out the bolt carrier. This can be a real plus when the smaller caliber is suddenly needed, as might be the case while hunting or when in a survival situation. It does, however, take forever to get off a second small caliber shot (though a second shot in 5.56mm is easy) unless you have more than one adapter. Even then, they have to be individually fed by hand into the chamber.

Single-shot adapters are available from Harry Owen (Box 5337, Hacienda Heights, CA 91745, 213/968-5806) for $24.95 each in stainless steel or $16.95 each in blue steel. The .22 LR in .223 Remington (5.56mm) will allow you to fire .22 Short, .22 Long, .22 Long Rifle, or .22 CB Caps. Another adapter is needed in order to fire the .22 Magnum in .223 Remington.

CCI CB Caps are especially useful for nearly silent shots when and if such are needed. The single-shot adapters lend themselves well to this, but the .22 LR conversion kits mentioned above also allow the use of CB Caps if the shooter cycles the action by hand.

9mm LUGER CONVERSION UNITS

At the time of this writing, several companies are offering a 9mm Luger conversion unit for the AR-15. Probably the best known of these is that offered by Frankford Arsenal (a private company). This kit consists of a 16-inch 9mm barrel mated to

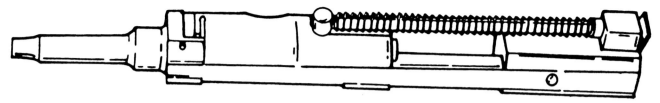

M261 .22 LR conversion insert

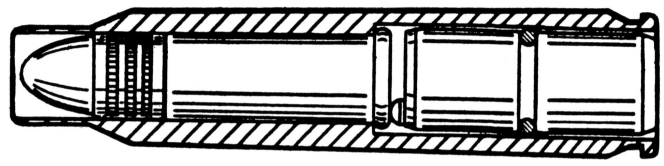

Harry Owen .22 LR/.223 Remington adapter

a standard receiver, a special blow-back operated bolt, and a magazine well insert that allows a Sten magazine to be used with the rifle. This unit is made to fit into a standard receiver with front and rear push pins.

The AR-15 buffer system becomes the recoil spring for the conversion unit, and a new magazine release lever extends out the lower receiver just behind the Sten magazine.

To get the 9mm conversion unit ready to use, it is necessary to first push out the front and rear push pins, take off the standard upper receiver/bolt/barrel, and put the 9mm conversion unit in place. Place the new 9mm magazine "well" inside the standard magazine well of the lower receiver, close the upper and lower receivers together, push in the two pins on the receiver, and place a full 9mm magazine into the well. The unit is then ready to be cocked and fired. The entire changeover takes only as long as it does to describe the steps once you get used to doing it.

Except for the magazine release, the rifle is operated in the standard manner. Safety, cocking, aiming, and handling are all the same as in the 5.56mm rifle. If you have an automatic version AR-15, the 9mm conversion will also fire in full auto or semi according to your choice.

The Frankford Arsenal 9mm conversion unit is also available with a 10-inch barrel or with either internal or external suppressors if you are licensed to receive such a weapon. The internal suppressor is interesting in that the unit occupies the inside of the handguard so that the rifle appears to be a standard AR-15 carbine with telescoping stock.

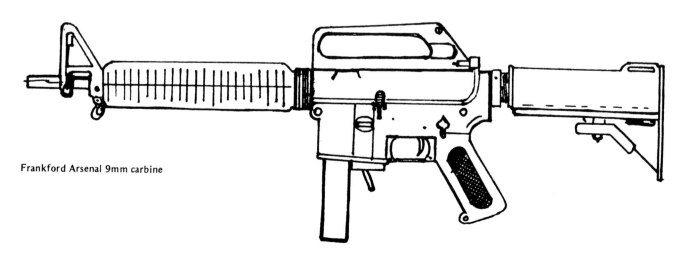

Frankford Arsenal 9mm carbine

The 9mm conversion unit is available from Frankford Arsenal, Inc. (1047 Northeast 43rd Ct., Fort Lauderdale, FL 33334, 305/566-8690) for $350 ($25 more for heavy barrels). Suppressors, short barrels, etc., are available on special order.

Frankford Arsenal also offers a complete 9mm carbine, the "M-9," for $703.92. This rifle will accept the military AR-15 upper receiver (with front and rear push pin holes) for those who wish to start with a 9mm carbine and later add the 5.56mm option to it.

A slightly more radical 9mm and/or .45 ACP adapter has been developed by Bill Holmes. It consists of a completely different design for the upper receiver (rather than using a standard upper receiver like the Frankford Arsenal unit does.)

The Holmes' upper receiver is squared off to allow the use of an integral pair of recoil springs to cycle the large bolt of the conversion unit (in a manner similar to operation of an AR-180 rifle). The receiver design does away with the need for the slip ring, weld spring, etc. Instead, the barrel screws into the front of the receiver, the one-piece handguard goes over the barrel (no gas tube is needed since the action is blow-back), and a front sight base is placed on the barrel. A nut is slipped on in order to hold the barrel and handguard in place. (The handguard will probably be narrower than the standard round AR-15 grips; Holmes, who has designed a number of rifles, pistols, and shotguns, feels that this makes for an easier-to-handle weapon.)

The new squared-off, steel upper receiver on the Holmes conversion kit is a lot cheaper to manufacture and offers some unique features, including a cocking handle, similar to that of the AR-10 and early AR-15s, inside the carrying handle. Probably the most interesting feature is that it is possible to remove the barrel and replace it with a .45 ACP barrel. This, along with a new magazine, magazine adapter, and bolt allows the owner of one of the upper receivers to shoot two calibers in his carbine for just a little more money than one conversion receiver costs.

Since the recoil springs are in the upper receiver, the regular AR-15 buffer and buffer tube are not needed in order for the unit to function. This raises some interesting possibilities if the owner of the carbine wishes to only fire the pistol rounds: the buffer tube/stock can be taken off and replaced with a folding stock.

If a person purchases a shorter barrel and completely removes the stock and buffer from the lower receiver, it would also be possible to create an "assault pistol" in the K-99 style.

At any rate, the Holmes conversion unit and the ability to change from 9mm Luger to .45 ACP will make the unit less expensive than the Frankford Arsenal version. We suggest you write or call Holmes Firearms Co. (Rt. 6, Box 242, Fayetteville, AR 72701, 501/521-8958) for more information in regard to price and available options for the 9mm and .45 ACP conversion units.

A third company has worked out an upper receiver conversion for AR-15 rifles. This conversion

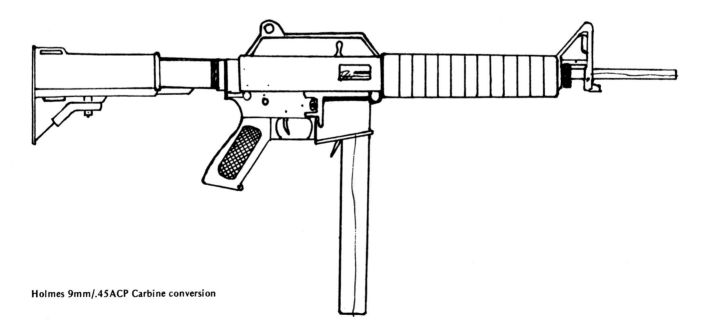

Holmes 9mm/.45ACP Carbine conversion

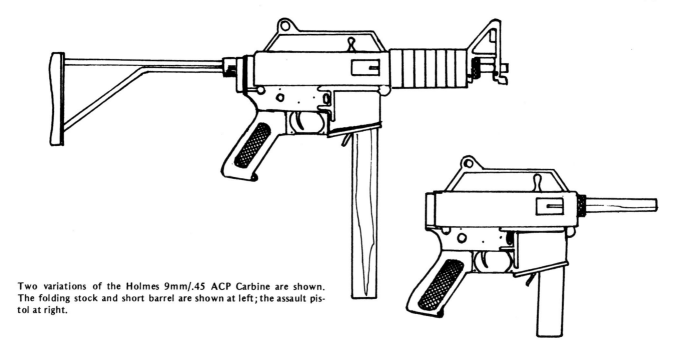

Two variations of the Holmes 9mm/.45 ACP Carbine are shown. The folding stock and short barrel are shown at left; the assault pistol at right.

is being marketed by A.I.I., Inc. (P.O. Box 26483, Prescott Valley, AZ 86312) under the product name of A.R.C. Weapons System. At the time of this writing, the conversion units are available in 9mm Luger and .45 ACP. A .22LR version is being developed and a .380 Auto chambering may be available via a barrel change of the 9mm system.

The A.R.C. system is similar to a cross between a MAC and a Sten. It has a Sten magazine feed into it from the safety side of the rifle, which does away with the need for a magazine adapter in the well of the lower receiver of the AR-15. The new upper receiver also has a squared-off body with low-slung sights similar to that of the MAC-10.

Because the magazine feeds from the side, it is possible to exchange the upper receivers and be ready to go with the new caliber! This makes it possible to switch from one caliber to another considerably more quickly than with the other two 9mm conversion systems described above.

The steel unit has a modified "V" rear sight with a peephole at the lower tip of the "V" and a post as the front sight. Sight pictures can then be gained rapidly (the system is designed with close-quarter combat in mind).

The A.R.C. operates on a single-spring, blowback system with a floating firing pin. It can be used on either a semiauto or automatic version of any AR-15. Like the Holmes design, the standard AR-15 stock is not needed on the A.R.C. in order to operate.

The barrel is available in 4-, 5-, and 16-1/2 inch versions, with the shorter versions being threaded for use with a Sionics-type silencer (which can be quite effective with pistol bullets).

A short-barreled A.R.C. pistol is also available. It is made so that the lower receiver of the pistol cannot be easily converted into a rifle receiver, thus getting around BATF restrictions.

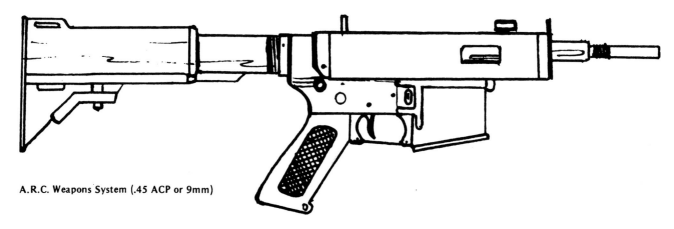

A.R.C. Weapons System (.45 ACP or 9mm)

The complete upper receiver is available in the caliber of your choice for $275, with the cost of barrels ranging from $40 to $120 (depending on their lengths). Spare magazines are available. A complete A.R.C. upper/lower receiver weapon is also available for $400.

The fourth entry into the pistol-caliber conversion race is the SGW CAR-9 barrel assembly, which is similar to the Frankford arsenal conversion. The SGW kit comes with the barrel, handguards, forward assist, upper receiver, etc., so that it can be added to an existing AR-15 rifle with a front push pin in a matter of seconds. The kit (available from SGW, 624 Old Pacific Hwy., S.E., Olympia, WA 98503), comes complete with a 34-round magazine for $359.

While working on this book, I contacted the BATF to check into the legal ramifications of the pistol versions of the A.R.C. and Holmes AR-15 conversion kits. It is a little up in the air! One of the BATF people felt it would be legal *if* the conversion and lower receiver had never been used as a rifle (in which case the pistol would be considered a shortened rifle), while another said that the receivers were manufactured for rifles and any barrel therefore under sixteen inches mounted on any receiver would be illegal without a special permit and prior approval.

In fact, the pistol versions probably are legal *if* the lower receiver is brand new and never used as part of a rifle, and *if* you cannot mount the short unit on an AR-15 rifle. (This might be achieved by welding a rod through the front push pin area of the pistol so that the receivers cannot be taken apart.) Some enterprising company, no doubt, will also start to manufacture a "pistol" receiver for these units if they become popular, so that it will be possible to get around the restrictions without legal complications.

Steven E. Higgins, director of the Bureau of Alcohol, Tobacco, and Firearms, wrote in the March 1984 issue of the *American Rifleman*:

Individuals who wish to modify a rifle to construct a firearm having an overall length of less than 26 inches or a barrel length of less than 16 inches must first submit an application to make and register a firearm, pay a $200 making tax, and receive approval of the application, prior to making the firearm. However, the provisions of the NFA [National Firearms Act] should not apply to pistols made from new, unused rifle-type actions which have never been assembled or barreled

as a rifle.

So, at this point in time, it would be possible to construct such a pistol but, to be on the safe side, you should write to BATF to be sure it is still legal. Be sure to get it in writing, not via a phone call. You should also get a permit before you order or start work on a short-barrel conversion kit if it will be capable of being mounted on a regular AR-15.

Of course, you should ask yourself whether you even need such a pistol. It would be big, awkward, and hard to aim without a shoulder stock. It would certainly be more of a collector's piece or adult toy than it would be a real self-defense weapon, unless you build an automatic version. In such case, you might as well add a folding stock since the BATF tax is the same whether the weapon has too short a barrel or the legal 16-inch one if it is registered as a machine gun.

Interestingly, it is possible, in a pinch, to make 9mm brass from old 5.56mm brass. To do this, it is necessary to cut the mouth of the 5.56mm empty down to the proper length for the 9mm, full-length, resize it in a 9mm Luger resizing die, trim the length again, and ream out the inside of the brass to .351 inches. Loaded with a 9mm bullet, the round will function well in most 9mm actions. The only thing to be careful of is to not go toward maximum loads with the brass since the head dimension of the 5.56mm cartridge is smaller than that of the regular 9mm. Be sure to test the rounds in your carbine to be sure the extractor/ejector will work with the round and that they'll feed into the chamber properly.

THE 6mm-.223

At one time barrels chambered for the 6mm-.223 Remington Wildcat round were available for the AR-15. Mounted on a receiver, a standard AR-15 lower and magazines would work with the barrel to enable the shooter to fire the Wildcat round.

The 6mm-.223 or 6x45mm was developed in the mid-Sixties, and is made by necking up a standard 5.56mm brass and adding a 70- to 100-grain bullet to it along with the proper powder charge and primer. The Wildcat does enable the shooter to fire a heavier, though slower, bullet which is more stable than the original FMJ 55-grain bullet of the 5.56mm in crosswinds. However, the new, better-designed 5.56mm bullets are doing away with this advantage. Another benefit of the 6mm-.223 is that it has a much quieter report and will work with standard AR-15 magazines.

If you should purchase a barrel, be sure to examine it to be sure it is in good shape (it might be pretty old), and try to get a 1-in-10 inch twist since it will stabilize bullets from 70 to 100 grains. The 1-in-12 twist is good only for bullets lighter than 85 grains, and the 1-in-14 twist is capable of stabilizing bullets of only 75 grains or less (which puts you into the ballpark of some of the new .223/5.56mm bullets).

OTHER CONVERSIONS AND MODIFICATIONS

Currently, several interesting conversions for the AR-15 and the AR-15 shorty carbine are being offered by Jonathan Arthur Ciener (6850 Riveredge Rd., Titusville, FL 32780, 305/268-1921).

Converting the AR-15 to fire from a belt can be done on the owner's rifle (provided it has all Colt parts) for $865. A whole-new rifle can be converted to belt feed and is available for $1,295 for the semiauto rifle, $1,338 for the semiauto carbine, $1,445 for a selective fire rifle, and $1,448 for a selective-fire carbine.

Arthur Ciener is a Remington 870 (pump) or 1100 (semiauto) shotgun which mounts under the barrel of the AR-15 by fastening to the bayonet lug in front and to a modified front push pin at its rear.

Although this modification looks wild, it is a little hard to imagine when it might be of use other than in extremely specialized situations when tear gas rounds or the like might be launched from the shotgun. The shotgun/AR-15 combination makes an extremely heavy load since the legal barrel length is eighteen inches for shotguns (two inches longer than the legal rifle barrel length). This creates a rather awkward combination, though the barrel can be additionally shortened with a special federal permit.

A shooter should consider long and hard before he goes with this conversion. Is it necessary? Would an automatic or burst mode be more effective than a shotgun?

If the "Over/Under" option is needed, it is available from Jonathan Arthur Ciener for $250 if you supply the shotgun (to be modified for

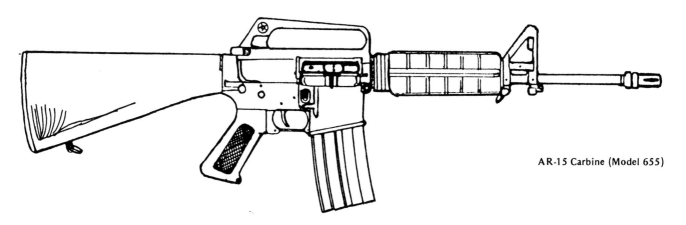

AR-15 Carbine (Model 655)

During recoil, the belt feed mechanism uses an angled camming surface on the bolt carrier to run a "seesaw" double fork bracket, which positions the round from the belt for proper pickup as the bolt is forced forward by the buffer spring. The weapon fires from a closed bolt (good for accuracy, not so good for heat buildup) and has an ammo carrier which will hold one hundred belted rounds of ammo and catch the ejected links of the disintegrating belt so they can be reused. The rifle can be broken open by pushing out the rear push pin, and the feed mechanism can be removed so that the AR-15 can use regular magazines.

The second conversion offered by Jonathan

either the semiauto or automatic version of the AR-15), or used shotguns can be purchased from Jonathan Arthur Ciener with the modifications already made for $435 for the 870 pump or $675 for the semiauto 1100.

An interesting modification of the gas system for the AR-15 rifle was available for a short time from Rhino International Corporation (215 Shadeland Ave., Lansdowne, PA 19050). This conversion makes the AR-15 operate with an impinging rod (similar to the AR-180 or FN FAL) so that the gas tube is eliminated. The gas and powder residue are then no longer channeled into the bolt of the rifle.

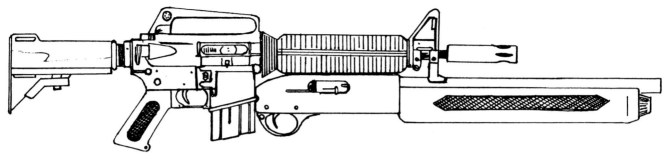

Carbine with Remington 1100 semiauto shotgun

The Rhino system, as the name implies, modifies the front sight base so that it has a gas tube into which the gas piston and its rod fit. When the rifle is fired, the gas comes out the regular gas hole in the barrel and pushes the piston and rod backward. The tail end of the rod propels the special key and bolt carrier backward, and the action cycles in the usual manner. Since the rod is the same size as the gas tube, no modification of the receiver or handguards is necessary.

In order to use a small, hardened steel operating rod, it is necessary to create a miniature buffer system, which is incorporated inside the replacement key. This buffering system consists of a series of neoprene and steel discs. Along with the discs is a short rod with a large groove in its center which holds the discs in place when it is trapped in the upper end of the key by a "dimple" peened into the key wall. When the rod is hurled back by the gas, it collides with the short rod and the plastic and steel discs. The shock of the collision is absorbed by the spacers.

The gas cylinder with the Rhino modification has an adjustable screw mounted on the front which allows the shooter to adjust the cycling rate of the weapon. This allows the user to change the rifle's rate of fire according to the ammuni-

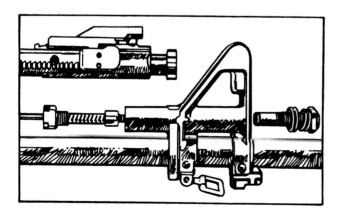

Rhino gas conversion system for the AR-15

tion used or lower the cyclic rate with automatic fire, thereby conserving ammunition when firing full auto. (The manufacturer claims a savings of two-thirds is possible.)

While this modification is not really necessary with new, clean ammunition or with the new automatic AR-15s with the three-round burst control, it apparently works and might really be of use under some circumstances if a modified barrel can be located or fabricated.

Another do-it-yourself modication involves making a superaccurate sniper rifle. To do this, it is necessary to get a heavy barrel with a fast twist, along with the gas tube, handguards, flash suppressor, etc. For the best accuracy, a scope mount, quality scope, and perhaps a cheek rest should be added. (See the accessory section for more information.)

It is possible to build a complete sniper rifle. Or, you can mount the barrel and its parts on an upper receiver, which can then quickly replace the standard upper receiver on your AR-15. This practically gives you, in effect, two rifles.

Standard handguards will fit on the extended barrels since the distance from the receiver to the front sight post is the same as on a regular rifle; the extra length is beyond the front sight base. Since the barrel on an AR-15 can flex slightly (though this is not as much of a problem with the heavy barrel), a match handguard might be desirable for the ultimate accuracy.

The steel handguard is a solid piece which screws around the receiver of the rifle so that the barrel is free-floating. This means that a bipod mounted on the handguard or a tight sling won't cause the barrel to bend. The shots will all remain on the same zero regardless from which position you fire.

The steel match handguard is heavy: two pounds. Its weight, coupled with the heavy barrel, can make the sniper rifle weigh eleven or twelve pounds. That is quite heavy to anyone who is used

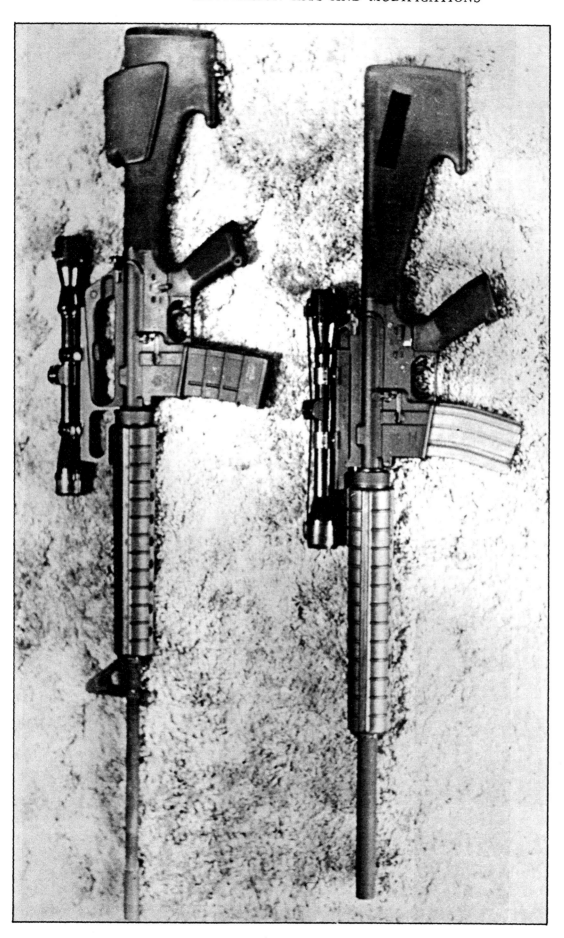

Two heavy-barreled sniper rifles the author has built are shown here.

The author has built this 24-inch, heavy-barreled sniper rifle.

to carrying an AR-15 (though normal for those poor souls who insist on HK-9s or M14s). The weight does have an advantage: When you shoot the rifle, it eats up recoil to the point where it is almost nonexistent. Currently, match handguards are available from Rock Island Armory (420 West Main St., Geneseo, Il. 61254, 309/944-2109) for $30.

Assembly of the rifle is basically the same as that for a normal AR-15 rifle (detailed elsewhere). Unless you will be shooting the new NATO tracer rounds, the 1-in-9 twist is probably the best bet. If you are using the rifle for combat and don't need to penetrate barricades or the like, the soft or hollow point bullets would be the most lethal. (Solid points won't tumble excessively on impact with this fast of a twist.)

For extra velocity—which can make a difference with long-range shooting—try to get a 24-inch barrel.

Excellent heavy barrels with a 1-in-9 twist are currently being offered by L. L. Baston Co. (Box 1995, El Dorado, AR 71730, 800/643-1564) for $100 for the 24-inch barrel or $95 for the 20-inch barrel. If you wish to create a sniper rifle conversion or need extra accuracy, these barrels are a good place to start.

You could, in theory, improve the accuracy of the metal sights by mounting a new front sight toward the muzzle of the barrel, but this is a lot of bother and requires machine work to really be good. A better bet would be to get a scope.

Due to the weight of the rifle, it is hard to imagine using it for anything other than sniper/countersniper work. As such, it would need a scope and the sights would probably not be needed except if the scope failed. In such a case, the distance would probably be too great to rely on the metal sights. Therefore, it might be wise to do away with the sights and use the scope.

The Quicksight System, a very useful upper receiver modification, is useful on such a sniper rifle. It is a standard receiver on which the carrying handle/rear sight has been milled off and a standard scope base securely placed on top of the receiver. This is a big improvement if a scope needs to be used since it does away with the need of a cheek pad, and it keeps the scope from getting caught in brush when you are carrying the rifle.

Though the scope mounted on the Quicksight System receiver will work without the removal of the front sight post, taking the front sight off will make a brighter "picture" during poor light conditions.

To remove the front sight base, use a hacksaw to cut through the mount above the gas tube area. Use a grinder and file to smooth and even up the cut. Touch-up gun blue will allow you to get a nice dark finish so that the bright exposed metal no longer shows.

If you have two rifles or two receivers which you exchange, the Quicksight will allow you to use the same cheek weld and look at the same point when sighting in either the scope on the sniper rifle or the metal sights on the standard rifle. These pluses make it possible to quickly switch from one rifle to the other without changing mental gears.

The Quicksight System is available from Long (no relation to the author) Engineering (4340 Eaglemere Ct., SE, Cedar Rapids, IA 52403) for $149. The company will also modify your own upper receiver to the Quicksight configuration if you so desire.

Because of the weight of the heavy-barreled sniper rifles, flash suppressors which compensate for muzzle climb are not called for unless you will be using the rifle in an automatic mode (which would probably not be wise in a sniper weapon!).

The standard military surplus flash suppressor works just fine. The extended barrels are also available without flash suppressor threads, but I am somewhat leery of having the muzzle exposed; one ding could destroy the barrel's accuracy.

How accurate is the sniper rifle? It depends on your abilities, the ammo, and barrel. Typically, if everything is working at its best—including you—you should be able to stay within one to two minutes of angle (the standard AR-15 with metal sights is capable of 2.5 to 3 m.o.a.). That means that you will be able to shoot groups of 1 to 2 inches (and possibly less) at 100 yards. This accuracy should continue on out to 500 or 600 yards, at which point you could expect a 5- to 12-inch groups (and maybe better if you and everything else are up to it). For best results, you need to experiment with powder, cartridge, and bullet combinations. Many shooters find that the small 52-grain hollow points and the large 62-, 63-, and 69-grain bullets are much more accurate than the 55-grain full metal jacket.

Conversion units offer a lot of advantages to owners of AR-15 rifles, and any owner of an AR-15 rifle should give some serious thought to purchasing one or more conversion kits.

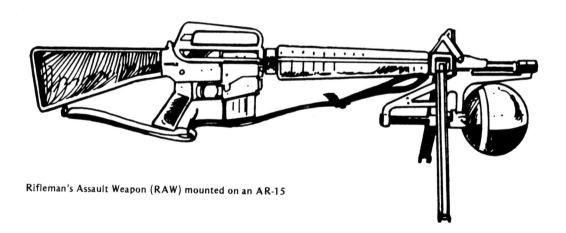

Rifleman's Assault Weapon (RAW) mounted on an AR-15

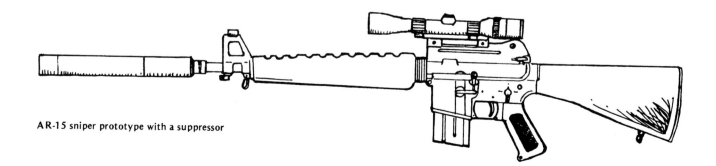

AR-15 sniper prototype with a suppressor

12. Grenade Launchers

One military trend of the twentieth century is to increase the firepower and capabilities of the infantryman. One route taken has been through the use of small assault rifles; another, through the use of grenades or other explosive devices which could be delivered by a foot soldier.

The hand grenade is very limited in both its range and the average soldier's throwing abilities. Since the throwing range is limited, the grenade, of course, cannot be lethal beyond the point to which it is thrown. To overcome these drawbacks, a number of weapons have been developed to better deliver such small explosive charges.

RIFLE GRENADES

During World War I, troops in the trenches often were issued a rifle grenade which was placed over the muzzle of a rifle barrel. The grenade was launched by gas from a blank, while the stock of the rifle was held against the ground in mortar fashion. The drawback with these rifle grenades was that the charge needed to send a grenade to a target of any great distance created excessive recoil; grenades therefore could not be fired from the shoulder, and a steady diet of grenades soon turned a rifle into a broken pile of junk.

One step toward a better rifle grenade was the "Mills Bomb" Mark 2 which was developed for the M-1 Garand rifle. It was mounted on the rifle barrel with a special adapter (the M6 launcher), and the rocket-looking grenade was then propelled from the weapon by firing a blank, just as in the World War I version. Regular ball ammunition could not be used to propel the rifle grenade, and the rifle could not fire ball ammunition with the launcher in place.

Though the rifle grenade was better than the hand grenade, it was very difficult to aim at dis-

tant targets. The rifleman had his weapon turned into a single-shot device for all practical purposes when he was set up to fire a rifle grenade.

Nevertheless, the rifle grenade did give a needed option to the rifleman. By the time the U.S. military was using the M14 rifle, a wealth of different styles and types of rifle grenades was available (including the M1A2 grenade projection adapter, which allowed regular hand grenades to be mounted and fired as rifle grenades). The grenade inventory included the M31 antitank grenade (capable of defeating ten inches of armor or twenty inches of reinforced concrete), smoke grenades (available for producing several colors of smoke for marking and signal purposes), star clusters, white phosphorus, and practice grenades.

Most of these grenades can be fired from the AR-15, since its flash suppressor was designed with this in mind. Currently the most common "Energa" rifle grenade can be fired from any AR-15 model except those with extra long or short barrels. The grenade is propelled from the rifle with a special blank cartridge (just like earlier rifle grenades were) and has a maximum range of 350 meters.

Generally the rifle grenade has fallen into disuse because it requires a special range finder to be accurate at long ranges, and it ties up the rifle with a special blank cartridge. The grenade must be launched with the blank cartridge before the rifle can be brought into play, or it must be removed and a regular round chambered.

Small hand-held launchers ("palm slaps") have been developed which can be used to rocket-launch pyrotechnic signals, smoke, and flares. These operate as well as the rifle devices, but with a lot less hassle. Despite the drawbacks, rifle grenades are

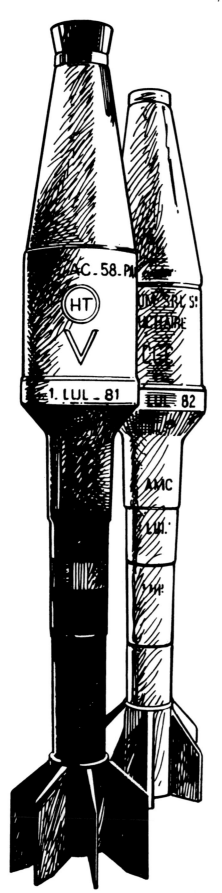

Luchaire rifle grenades with integral bullet traps

still seen from time to time around the world, and a number of rifle grenade devices are on the commercial market (except for the really dangerous explosive ones which are generally illegal to own without a special "destructive devices" federal permit).

Most common on the commercial market are parachute flares and practice grenades, and these will work with the AR-15. One company that has a good supply of these products, along with special blanks and a special grenade retainer spring to keep the device secured on the rifle (so it does not fall off when you are running) is Phoenix Systems (P.O. Box 3339, Evergreen, CO 80439, 303/674-2653).

It should be noted that the recoil from many rifle grenades is greater than that produced by the recoil of a cartridge. Whenever possible the rifle stock should be placed against the ground, while you keep your head clear of the grenade. A steady diet of rifle grenades is a bit hard on the rifle as well. Do not try to use ordinary blanks to launch rifle grenades, since the grenade will fall only a few feet from you, producing results you had not planned on!

THE M79 GRENADE LAUNCHER

Because the rifle grenade was awkward to use, tied up a rifle, and did not have as great a range as might be hoped for, research was conducted to create a better way to launch grenade-type devices via a special weapon designed specifically for that purpose.

The outcome was the M79 grenade launcher. The weapon was very similar to a single shot shotgun, but it had a 40mm caliber. (The device was probably an outgrowth of the tear gas grenade launchers used by police forces beginning in the Thirties.)

The M79 was successful, but it had several drawbacks: Many of the rounds were not armed until they had traveled for 30 yards (which often meant the user of an M79 was out of luck in combat at closer ranges), and the weapon was only single-shot.

To overcome the shortcomings of the M79, a special 12-gauge adapter was made to allow it to fire shotgun shells. This change alleviated the problems originally encountered with the weapon in close combat in a one-shot sort of way, but made it impossible to quickly fire regular M79 rounds. A special round was made which allowed the firing of buckshot; this made it possible to fire other M79 ammunition without pulling out an adapter, but the round was not much better than a regular shotgun shell and the cartridge was much larger. As a result, not much ammunition could be carried. Multibarrel grenade launchers with 3-round, 40mm launchers and a flechette barrel were tried but proved to be too heavy. (These were produced by Harrington & Richardson and Winchester as SPIW prototypes.)

THE XM148 AND M203

Perhaps inspired by field troops who were cutting down the stocks of M79s and wiring them to AR-15s, Colt started working on the XM148 grenade launcher in the late Sixties. The XM148 was to be attached to the underside of an AR-15. The M79 weighed 6.45 pounds loaded, while an AR-15 with the launcher only weighed a little over 9 pounds. The AR-15 and launcher increased the ability of the user to protect himself with rifle fire should the need arise.

The XM148 made its way to Vietnam, but it got the reputation of being overly complicated and sometimes unreliable. The M203 was consequently developed and standardized, and it was accepted in mid-1969 by the Army.

The M203 is capable of using all the rounds developed for the M79, thereby making it a very versatile weapon, even though it is only able to fire grenades single-shot. Maximum range for the M203 is 400 yards with a 30-yard arming range for some of the rounds.

Rounds for the M203 include High Explosive (HE), various colors of smoke (for ground use or on parachute), CS gas, flares of various colors for signaling and for illumination, multiple projectile rounds, and practice rounds. Most rounds weigh approximately eight ounces.

The rounds used in the grenade launcher have a high-pressure charge which is ignited in a small container in the base of the cartridge. This container ruptures and the pressure drops as the gas moves into a larger chamber and pushes the grenade out of the tube, leaving an empty casing behind in the chamber of the launcher. This roundabout method of launching the grenade makes the recoil much lighter than a conventional round the size of a 40mm grenade.

FRENCH RIFLE GRENADES

Although the United States, along with most other countries, has pretty well abandoned the use of rifle grenades in favor of grenade launchers, the French military is a notable exception to this trend.

The French rifle grenades will work on the AR-15 since they were designed with such use in mind. These grenades have a bullet trap in the base of the grenade which makes it possible to fire them with regular rounds rather than having to chamber a special blank before it can be fired. This change makes the French rifle grenade much easier to use since it can be quickly placed on the firearm's barrel and fired or removed if normal use of the rifle is required.

The Luchaire Company is currently manufacturing the grenades; they claim that with a little training a soldier using their special range finder can easily place the grenades through a window-sized opening at 150 meters. The company is manufacturing antiarmor rounds capable of defeating the side or rear of tank armor (250-300mm penetration of rolled homogeneous armor) and antipersonnel grenades with 60mm armor penetration and a lethal radius of 11 meters.

At the time of this writing all rifle grenades in production are being sold to the French military, but the U.S. 9th Infantry Division at Fort Lewis, Washington, and the U.S. Army Infantry Center at Fort Benning, Georgia, are both looking into the practicality of the rifle grenades. Luchaire claims the grenades could be manufactured in the United States at about $60 each for the armor-piercing or antipersonnel versions and slightly less for smoke or flare grenades. These are undeniably bargain-basement prices these days for weapons.

THE RUBBERHEAD

An interesting variation on the rifle grenade is currently being marketed by the Israel Product Research Company. This grenade is not designed for combat, but rather for riot/crowd control. Called the "Rubberhead Projectile M-809," it has a large rubber ball atop a conventional rifle grenade. The grenade can be used to hit a target from 160 yards and gives a knock-out punch, which is less apt to injure the person than plastic or rubber bullets. The projectile is also filled with 80 grains of tear gas so that others in a crowd will be affected by the grenade.

The Rubberhead has to be fired with a special blank, but this is probably less of a problem in crowd control than it would be in combat. (Sometimes, though, it is difficult to distinguish the difference if you are in the thick of it!)

THE RIFLEMAN'S ASSAULT WEAPON

Another innovation along the lines of the rifle grenade is the Rifleman's Assault Weapon (RAW). It looks somewhat like the rifle grenade, but it is in fact a rocket. The 6-pound RAW is mounted on the barrel of an AR-15 (its base locks on the high front sight and the bayonet lug). It is fired when a regular bullet is caught in an integral bullet trap and the impact of the bullet's gas sets off a percussion igniter, thereby starting up the rocket motor. Unlike grenades which have a very steep trajectory (and are therefore hard to aim at long ranges), the RAW has a flat trajectory created by a downward as well as forward thrust. It is thus kept parallel to the aim of fire for 200 yards (it is capable of a 2,000-yard shot if a sighting device is used). In addition to the downward and rearward rocket exhaust, two tubes divert another portion of the rocket's thrust to create a stabilizing spin for the projectile.

In addition to the welcome use of a bullet trap and the ability to aim the RAW with regular rifle sights, the rocket does not create any recoil so the firing strain on both the rifle and the rifleman no longer exists. The rocket has a very low noise signature and is, in fact, slightly quieter than a regular rifle blast. Since the RAW develops no back blast, it can be fired from indoors or from vehicles (unlike some tube-launched rockets).

Manufactured by Brunswick Corporation during 1980 for evaluation by the Advanced Systems Concept Office, U.S. Army Missile Command in Huntsville, Alabama, the RAW has explosives which could make it effective against armor, bunkers, walls, or vehicles *if* the weapon ever goes into production. Unfortunately, the RAW development seems to have been tabled in order to divert funds to tank and helicopter research, considered more urgent by the U.S. military. Only time will tell whether the RAW will ever be added to the infantryman's bag of tricks.

A rocket round may be developed which can be launched from the M203. The "plus" with such a development would be that the rocket round could have a larger payload while producing less recoil when fired from the M203.

Until—and if—RAW or M203 rockets reach the grunts in the field, it appears the M203 with its M79-style grenades will be used to give ground

troops the long-range grenade options often needed in combat.

A large number of M203 parts are currently making their way to the suplus market, except for the receiver/trigger group of the mechanism. The company which seems to have the parts in stock most often is Numrich Arms Corporation (West Hurley, NY 12491, 914/679-2417). Since the barrel is chambered for an explosive grenade as well as flares and smoke, it would be illegal to build a working M203 without BATF, state, and local approval.

OTHER DEVELOPMENTS

At the time of this writing, Frankford Arsenal, Inc. (1047 Northeast 43rd Ct., Fort Lauderdale, FL 33334, 305/566-8690) is in the process of tooling up to build commercial versions of the M203. They will be marketing at least two models, one chambered for the military 40mm round and another chambered for 37mm commercial flares. The 40mm version will be available only to persons who have the BATF, state, and local approval necessary to own one; the other version will be available to anyone since it won't be capable of firing anything other than flares.

Frankford Arsenal may also offer a 12-gauge adapter for the civilian flare version of the M203. It would be illegal to use the adapter in the flare gun without a BATF permit since the barrel of the launcher is only twelve inches long. With the adapter, the gun would be considered a shotgun.

Another route which might be taken by commercial manufacturers is to make a one-shot, black-powder grenade launcher look-alike. Since black-powder weapons are not restricted by caliber or barrel lengths, anyone who can normally obtain a firearm could own one legally. The M203-style weapon could be loaded with some of the black powder substitutes and would be a real short, legal—though awkward to reload—super-shotgun. Such a device would be more a collector's item or military reenactment prop rather than a weapon suitable for anything but the most limited of combat. (Perhaps, though, it would be sobering for a criminal who broke into a house and confronted a home owner who leveled such a weapon mounted on an AR-15 at him.)

OPERATING THE M203

The operation of the M203 is fairly simple and straightforward.

To load (or reload) the grenade launcher, the barrel latch (on the same side as the AR-15's selector) is released. The latch is about halfway forward to the top and side of the launcher's barrel.

When the latch is depressed, the barrel can be slid forward by grasping the ringed area of the barrel. Sliding the barrel forward will automatically cock the weapon, causing the extraction of any cartridge that is in the barrel since the round or cartridge will stay behind on the receiver (where the cartridge is held by the extractor).

With the barrel open, a new cartridge is inserted into the breech. The barrel is then pulled back toward the receiver, causing the barrel to lock up. The round will then become engaged by the extractor. The barrel latch will automatically lock the barrel in place.

Unless the weapon is to be fired immediately, the safety should be rotated backward, toward the trigger.

To fire, aim the weapon, push the safety forward (if the safety is engaged), and pull the trigger. This will release the spring-loaded firing pin which will ignite the primer, firing the round.

A large number of rounds are available for the M203. Probably the most useful in combat are the multiple projectile rounds (firing buckshot) and the high-explosive rounds.

The buckshot rounds (XM576E1 and XM576E2) have a range of only 50 yards for a good-hit probability; each round contains 27 pellets of 00 buckshot, making it roughly equivalent to a load of 12-gauge buckshot. HE (High Explosive) rounds have a maximum range of 400 meters with a casualty radius of 5 meters; the HE rounds are armed when they achieve a certain number of fast rotations. These rotations are made as the projectile spins after leaving the barrel, usually at 30 meters from the barrel.

There are two aiming systems for the M203. One sight is mounted on the top of the handguard just behind the rifle sight, while the other is a quadrant sight assembly that mounts on the AR-15's carrying handle.

The handguard sight is used in conjunction with the front rifle sight. It is folded up so that it is at right angles to the barrel, and the firer sights down the sight using the sight alignment markings (which are in hundreds of meters: the 1 being 100 meters, etc.). When the target is lined up with the correct range, the weapon is fired.

The quadrant sight is a bit slower and slightly more prone to getting hung up when the M203 is being lugged about. It is usually more accurate if you can judge ranges well.

To use the quadrant sight, estimate the range, and dial up the range on the elevation scale, which is marked in 25-meter increments. When the range has been set, sight through the rear sight of the quadrant sight using the front sight post at the front of the quadrant sight to line up on the target. When the target is aligned, fire.

Care should be taken not to get the front sight/elevation screw of the quadrant sight out of whack when carrying the weapon.

To mount an M203 launcher on a regular AR-15, remove the rifle's handguards from the barrel, and slide the M203 onto the barrel by placing the open, upper rear end of the M203 onto the middle portion of the barrel (where it is narrow enough for the M203 to fit over it). Once the slot of the M203 is on either side of the rifle's gas tube, push the launcher back toward the AR-15 barrel's chamber.

An insert is next placed between the forward end of the rifle's gas tube and its barrel (a short distance behind the rifle's front sight base so that it joins its counterpart/half-ring on the M203). The bracket is now placed over the gas tube and fastened to the top front of the M203 body. This will completely fasten the M203 in place.

If desired, the sling mount can be put on the right or left side of the barrel by placing it around the barrel inside either leg of the front sight base. The mount is held in place with two roll pins which should be drifted into position. (Many M203 users prefer the assault carrier sling arrangement described in the accessories section of this book.)

Finally the special M203 handguard, which protects the user's fingers and the rifle's gas tube, is slipped into place over the rifle gas tube and barrel, and the quadrant sight mounted if desired. The weapon is now ready for use.

Disassembly of the M203 is basically a reversal of the above procedure.

13. Automatic-Fire Conversions

It has been my experience that the less a person knows about combat, the more he feels the necessity to use an automatic weapon for self-defense. Automatic fire is sometimes effective, but it is most often very wasteful. Automatic fire requires a well-trained gunner to be effective; at small distances from an automatic weapon, there are a lot of "holes" between points where bullets are actually impacting. It is easy to miss a target after firing an entire magazine full auto from the hip! The winner of a gun battle is not the guy who makes the most noise, but the fellow who makes the first, telling hit.

The new 5.56mm is more controllable than older, larger calibers and comes close to being as controllable as a submachine gun, while also having a range rivaling the heavier machine guns. Consider the three-round burst modification and effective muzzle stabilizers, and it can all get confusing!

Effective machine gun fire seems increasingly rare on the battlefield since troops have less training with marksmanship and carry weapons which allow automatic fire.

If you do not have a nearly endless supply of ammunition, are not involved in house-to-house combat where targets are being engaged within thirty yards, and/or do not have a fixed position to defend, it is doubtful you really need an automatic AR-15. Most home owners or survivalists would certainly be better off with semiauto weapons since a rifle is too unwieldy and too powerful indoors (though a short carbine AR-15 with a pistol caliber conversion might make sense for indoor self-defense).

LEGAL CONSIDERATIONS

And there are legal problems in acquiring an automatic weapon. Many states do not even allow private citizens to own automatic weapons, though those who work for police departments or other "tax-free" organizations can shoot auto weapons held by the organization. Consequently, many small police departments have large arsenals of automatic weapons.

These state laws, coupled with the federal law, have gotten a lot of people into terrible trouble, often leading to the loss of firearms and even prison terms for the owner of the illegal weapons. Once the legal problems are settled, those guilty of owning an automatic weapon can no longer legally own *any* type of firearm since they are felons.

Anyone interested in having an automatic weapon should give a lot of thought to whether they really need and can afford an automatic weapon, and should go through the work of getting a legal weapon.

If you must have an automatic weapon, the first step is to contact your State Attorney to try to determine what the state laws are in regard to automatic weapons; you might also want to contact your state BATF agents to double-check. Always write to these officials and get letters outlining their position as to what they feel the law is. Officials often try to discourage private ownership of automatic weapons so, if both groups seem vague or give conflicting advice, contact a lawyer. (If you do run into problems later on, you can always verify that you tried to get the information pertaining to the legalities involved.) Do not check with your local police department to find out what the laws are in regard to automatic weapons, since police departments seem to be notoriously unaware of what the actual laws are in this regard and may give you erroneous information.

The easiest way to get an automatic weapon is to buy a legal one. Many of the companies listed in other chapters of this book offer automatic versions of the AR-15 that are very reliable and which often work better than those rifles that are modified at a later date to work in an automatic mode. If at all possible, you will usually have fewer problems if you get one of these weapons.

If you plan on converting a weapon, be sure to get all the paperwork and red tape out of the way *before* you start making any modifications to your weapon. If you convert your weapon before you have permission, you will have an illegal weapon. Once a weapon is illegal, it is always illegal, and there is no legal way to undo the damage once an AR-15 is converted illegally.

If you are currently able to own an automatic weapon, getting a federal license to own one will require having your fingerprints taken, some background investigation done, and payment of a one-time $200 tax to own the weapon. You may also be in for quite a wait. Do not delay in getting things started.

Once you are legally in the clear, you need to make a few decisions as to what type of modifications you wish to make to your AR-15. Generally, it is a good idea to go ahead and also get a short barrel on the rifle since you can do so without paying the extra $200 a shortened barrel requires. (You can get a short barrel and the automatic conversion for just $200 rather than paying $400 if you want one of each.) Having an upper receiver and regular length barrel will enable you to quickly switch over whenever needed.

WAYS NOT TO CONVERT THE AR-15

There are a number of ways to convert an AR-15 to auto fire, and some, of course, are much better than others. Most people assume that semiauto rifles are hard to convert to selective fire. Not so. It is relatively easy to do, and doing so on the AR-15 is no exception (though not as easy as some).

But not all selective-fire conversions are good.

Let us first look at some poor conversion methods so that you won't be tempted to try them. First, though, let me again give this warning: *It is against the law to alter a semiauto weapon to fire in a full-auto mode without the appropriate licenses from federal, state, and local governments. Severe penalties are prescribed for violations of these laws.* Be forewarned.

One way of converting an AR-15 to auto fire is to place an automatic rifle disconnector, safety selector, hammer, and trigger into a commercial semiauto AR-15. This will enable some slam fires when the rifle is placed in the auto mode. As mentioned elsewhere, this is nearly suicidal, since sooner or later one of the primers will fail to fully ignite and a bullet will be lodged in the barrel. This will be followed by another round which blows the whole thing—and you—up.

Pistol primers might give full ignition, but they will probably give you a fired round every time a round was chambered since the floating firing pin of the AR-15 is too rough for pistol primers. Certainly a weapon that starts firing full automatic on its own when you place a magazine in it and chamber a round is not ideal!

Another poor conversion is done as follows: The hammer is locked back and the safety placed into the safe position. Once this is done, the receiver halves are opened, and a striker (a rod which moves freely in the bolt carrier so that it can strike the rear of the firing pin) is placed in the bolt carrier. The user pulls back the carrier and locks it in place with the bolt hold-open latch, and places a full magazine into the rifle.

Releasing the bolt will cause the rifle to go full auto. It can only be controlled—somewhat—by using the bolt hold-open latch to start or stop it. This is about as bad as removing the disconnector and is not recommended. As we will see later, the striker is of use in creating an automatic version of the AR-15.

THE B.M.F. ACTIVATOR

One rather interesting device capable of being used for the conversion of the AR-15 to automatic fire is the B.M.F. Activator, available from B.M.F. Activator, Inc. (3705 Broadway, Houston, TX 77017) for $20. As it comes from the box, this device is legal to use on an AR-15. The B.M.F. Activator is a hand-turned cranking device that fastens to the trigger guard of the AR-15. When the crank is turned, a small tongue presses the trigger a number of times, thereby making the AR-15 operate like a modern-day Gatling gun.

The B.M.F. Activator enables many people to get around the legal limitations of a semiautomatic firearm while giving the capability of creating the number of shots per minute normally associated with an automatic weapon. It is not of "combat" quality, however, and serves only as a recreational device which can allow a shooter to have the thrill of firing an automatic weapon without the legal hassles.

The B.M.F. Activator does become the basis of an automatic weapon if a motor drive is added to it. As such, the Activator itself would probably be considered the "automatic weapon" by BATF, and you would therefore have to go through the legal red tape *before* it was converted. Serial numbers would have to be placed on it.

Though a motor-driven rifle is not suitable for combat, such a device might have some very limited use on a vehicle or plane. Unlike other conversions, this motor-driven unit can have an adjustable rate of fire which will vary according to the motor speed.

If such a setup were to be permanent, it would be necessary to replace some of the nylon parts of the B.M.F. Activator with metal parts. This is necessary since, though the nylon parts are quite tough, they warp over time if left connected to a rifle. Some sort of solenoid-operated safety would be needed, as would a remote aiming device (if

the rifle were to be aimed by some method other than aligning the vehicle with the target).

PRACTICAL CONVERSIONS

Probably the best route to take is to drill a hole through the lower receiver and install a standard auto sear. This requires very careful placement of the sear hole, but it creates the most reliable conversion. If you have a commercial semiauto AR-15, you will also need to purchase a military-style bolt carrier, as well as the automatic trigger, hammer, sear, and selector. (These parts are all readily available on the surplus market.)

A quick inspection of *Shotgun News* will generally turn up several companies which sell the auto sear. You will probably discover that several companies offer kits to convert a rifle to three-round-burst auto fire since these parts are becoming available on the surplus market. This is certainly an option to be considered. Once the red tape is done for converting your rifle to one mode or the other, you can switch to the other style if you so desire.

Rather than replace the semiauto bolt carrier, it

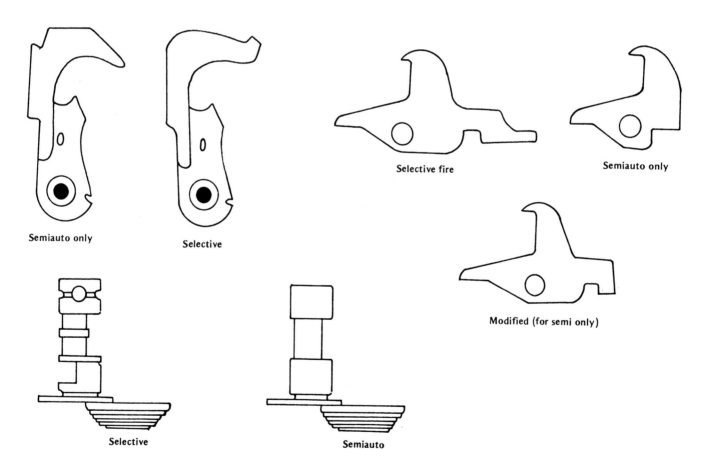

Semiauto only   Selective

Selective fire   Semiauto only

Modified (for semi only)

Selective   Semiauto

Comparison of selective-fire and semiauto parts

is possible to get a metal insert for the rear of the carrier. These generally work well provided they are carefully anchored in place and Loctite is applied to the screws so that the insert does not shake loose. For the utmost reliability, get a regular military bolt carrier.

One advantage of the AR-15 rifle is that if it is converted to selective fire by adding an auto sear to it, the lower receiver is considered the automatic weapon by BATF. You can thus substitute other upper receivers and conversions for the rifle and, in effect, have a number of automatic weapons for the price of one.

It is also possible to have your rifle altered for auto fire if you are leery of drilling a hole into your rifle's receiver. One good gunsmith who does this type of work is J. D. Farmer (1331 Hawthorne Ave., Smyrna, GA 30080, 404/433-8977).

If you do not wish to have a hole drilled in the AR-15's receiver, it is possible to convert it to selective fire by using the so-called "drop-in" auto sear and replacing the semiauto parts with the military-style bolt carrier, and military-style trigger, hammer, sear, and selector. Some rifles may

already have some of these parts; check before you spend money on parts you already have!

The drop-in sear generally works best with rifles that have matched receivers (i.e., an upper/lower that are made by the same manufacturer and do not require the special bolt/bushing needed to match the Sporter upper or lower to the military-style receivers.) Mismatched receivers occasionally close up with slightly different tolerances so that the drop-in sear may work well sometimes and not at all other times. (This action is caused by the camming action of the adapter offset screw needed to mate the two receivers together.)

The drop-in auto sear has been classed as an automatic weapon, and that is bad news to those who break the law, but good news to you if you need a legal automatic weapon. You can again pay the price for one automatic weapon while having several AR-15s that can be converted to selective fire by having all the essential parts except the drop-in auto sear.

The drop-in sear works just like the standard auto sear except that it has its own body, which contains the trip lever/sear and its spring and pin.

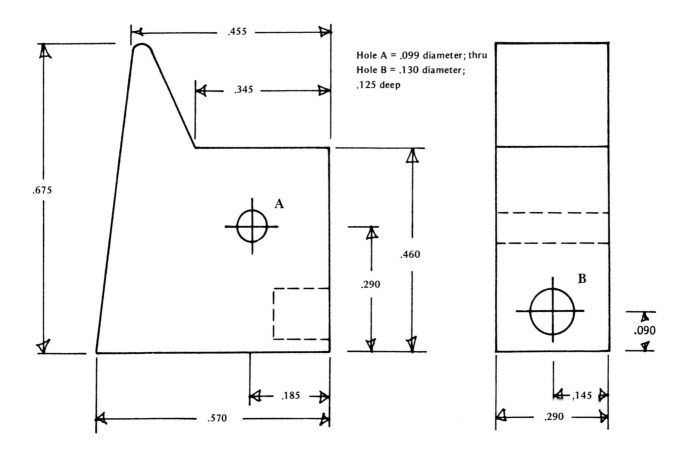

AR-15 sear

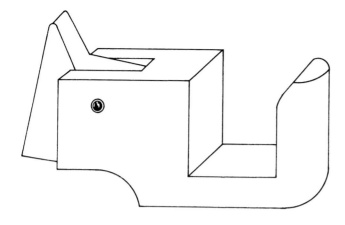

An assembled drop-in auto sear for the AR-15

The drop-in sear is available on the market, currently as parts since the $200 transfer tax must be paid before the assembled piece can be sold. Again, a quick search of *Shotgun News* will usually turn up several companies who offer the parts.

It is quite easy—and a lot cheaper—to make the drop-in sear yourself if you have had a little experience in metal work and have the patience to get all the measurements right. Patience is also required to do the file work involved in the final fitting of the drop-in sear to the AR-15.

The body of the drop-in sear housing can be aluminum, as can the sear itself, though a steel sear will give much, much better service. Closely follow the measurements shown in the illustration in this book. The spring for the sear should be somewhere in the neighborhood of 10 turns of number 18 music wire—about half an inch long—with an outward diameter of .125 inches. The pivot pin can be solid or a roll pin, and it should not drift out of place while still allowing the sear to move freely. Usually a roll pin .470 inch long and .093 inch in diameter will work well.

If you build a drop-in sear, be sure to make the housing so that it fits tightly into the space behind the safety selector. If there is any play, it probably will not work well consistently. Your best bet is to allow a little extra metal here and there, and file

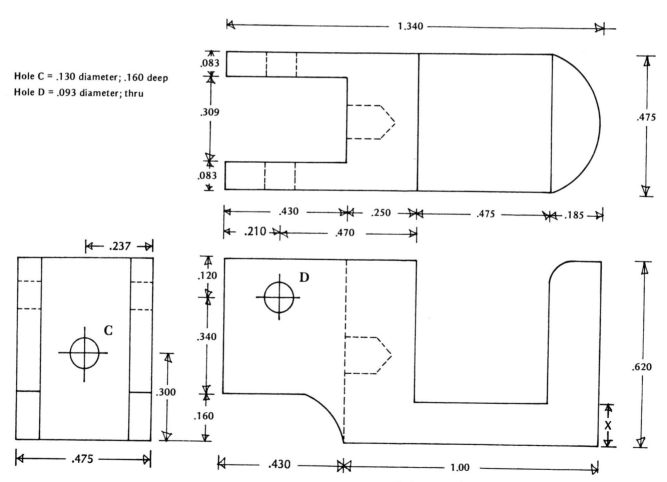

Hole C = .130 diameter; .160 deep
Hole D = .093 diameter; thru

X: determined by space below AR-15's rear push pin

The sear housing for the AR-15 is shown above.

fit it into place on your rifle. Cycle the assembly in the rifle by hand a number of times *before* trying it out with live ammunition.

When the auto sear is assembled, it must have a serial number placed on it since it is an "automatic weapon." According to the BATF regulations, the serial number should go on the largest part of the drop-in sear—in this case the sear housing.

BATF does not allow the serial numbers to be placed on the housing with an electric inscriber or the like since these are readily available to anyone who might steal your drop-in sear. It would be easy for such a person to alter the numbers. Rather, you must use a steel stamp to place the serial number on the drop-in auto sear.

This is not hard to do. The stamps are available from Brookstone Company (127 Vose Farm Rd., Peterborough, NH 03458, 603/924-9511) for $6.95 for a set of number characters one eighth of an inch high. To print a number, you merely place the stamp on the spot where the number goes and strike it with a hammer. The only trick is to keep the numbers more or less even. (A little practice on a piece of scrap metal will make a "pro" out of you.)

Another auto conversion is possible which will allow the rifle to be fired from an open bolt. If you use a heavy barrel with a bipod on your AR-15 and have a spare receiver/barrel assembly along with some 40-round magazines, you can create a reasonable SAW (Squad Automatic Weapon). This configuration certainly makes a very effective machine gun *if* you need one.

Drawbacks to open-bolt firing are somewhat reduced accuracy (since the bolt travels forward and locks shut as it fires, jarring the aiming point considerably on all but the most securely held rifles) and the lessened safety of the rifle. The greatest concern should most probably be the safety factor.

The safety problem arises because the rifle will fire when the bolt slams forward and a loaded magazine is in the well. This means that if the bolt is being held by the bolt hold-open lever, releasing it can cause the rifle to fire. Likewise, dropping a rifle with a closed bolt and a full magazine in place might cause the bolt to cycle back and then slam foward, firing a round. The same result would take place if you dropped a weapon with the bolt locked back and the safety on; the fall might jar the sear loose and cause the rifle to fire!

While this weapon is no more dangerous to the user than many other military machine guns, it is not nearly as safe as most commercial rifles. The user should therefore be aware of its shortcomings and be very, very careful when using it. The *open bolt modification can produce a weapon that is considerably less safe to use than the regular AR-15*. Therefore, if you try to convert an AR-15 to fire from an open bolt, be extra cautious when using the rifle. Remember that a bullet might come speeding down the barrel at practically any

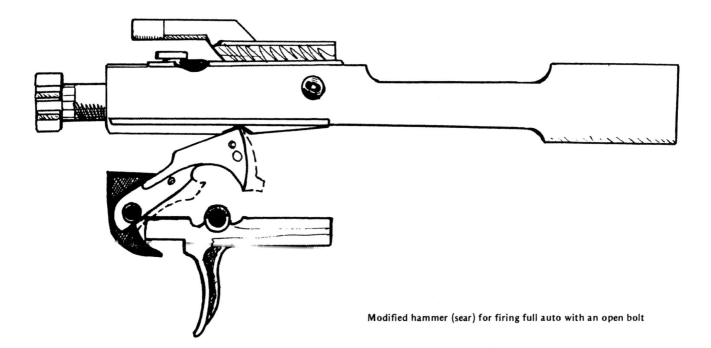

Modified hammer (sear) for firing full auto with an open bolt

moment. Do not point the rifle at anything you do not wish to shoot.

Though the safety factor is a drawback, the ease with which a rifle is converted is almost amazing. If you look at the open-bolt rifle diagram, you will see that all that is necessary to make this modification is to grind off the hammer's sear, trigger, and disconnector engagement spurs, and weld (or silver-solder) some metal extensions onto the hammer to limit its forward movement and to allow the lower front edge of the trigger to rotate the hammer (which will operate as a sear) back. It is necessary to grind a point on the top of the hammer and a slot in the bolt carrier into which the "hammer"—now a sear—is placed to engage the bolt and hold it back. (The hammer spring is placed on the hammer/sear in its normal position.)

The hardest part of this conversion is getting the correct notch cut in the bolt carrier; measurements are not given in this book because it is generally quite hard to keep the angles and curves of the hammer/sear exact. Consequently, it will be necessary for you to grind the notch in according to how you have shaped the hammer; take your time and get it right. An inverted "V" will probably give you the best results for the notch in the bolt.

Next, it is necessary to create a striker. This is a metal rod of .575 inch or less in diameter and 3 inches or less in length. The striker fits inside the rear half of the bolt carrier where it should move freely inside the carrier. The length and material of which the striker is made will determine to some extent the cyclic rate of the machine gun; you may wish to experiment with several different sizes of strikers. When you try them out, do so with only one round at a time to be sure the striker hits the firing pin with enough force to fully fire the cartridge's primer. A misfire may leave a bullet in the barrel, which can be disastrous on full auto!

For added safety, the bolt release should be removed. Though this sometimes creates a problem, it will make the rifle safer. No additional steps will be added to reloading the rifle since the charging lever would have to be retracted to bring the bolt back from the hold-open lever to the sear. Using a tracer round for the last one or two cartridges in a magazine will help you keep track of when you are coming to the end of a magazine.

The open-bolt conversion works in the following manner:

- The rifle should first be cleared to be sure it is empty.
- The charging lever should be pulled back until it locks.
- The safety is to be placed into the safe position.
- The bolt carrier will be held back by the sear so the charging lever is pushed back into place.
- A full magazine is positioned in the magazine well *after* the bolt is locked back.
- The safety is switched to fire position.
- The rifle is ready to be fired.

When the trigger is pulled, the trigger rotates the sear (modified hammer) back to release the bolt, which is then pushed forward by the buffer spring. As the bolt passes over the magazine, it strips off a cartridge and chambers it. When the bolt locks into the barrel extension, it allows the firing pin to come in contact with the primer of the cartridge. Since the striker is riding behind the firing pin, they will both have enough momentum to fire the cartridge primer. The bolt carrier is pushed back by the gas that is bled from the barrel, and the action is again locked back by the sear if the trigger has been released. If the trigger has not been released, the bolt carrier cycles again and fires another cartridge. This process continues until the trigger is released and the action is locked open or the magazine is emptied (if the trigger is still held back after the last shot, the action will close on the empty chamber).

Pluses with the open-bolt firing system include the ease with which an AR-15 can be converted and slower heat buildup of the system during sustained automatic fire.

A number of other auto conversions of the AR-15 are possible, but those mentioned above are the most practical for most users who desire auto fire.

An automatic AR-15 is not as effective in combat as many novice owners of the AR-15 seem to think. To maximize your fire, use short bursts and try to fire with a bipod from a fixed position if possible.

If you need an automatic weapon, it is not too hard to convert an AR-15 to selective fire provided you know what you are doing and do not mind wading through the red tape involved in obtaining a legal conversion before the work is actually done.

14. Troubleshooting

A large percentage of all problems with the AR-15 can be traced to the use of poor ammunition or an uncleaned weapon. Paying attention to these details can save you a lot of headache.

Most repair work should not be undertaken by anyone other than a skilled gunsmith. Firearms are dangerous—most especially a malfunctioning weapon. If your AR-15 is not functioning properly, take it to an expert repairman if at all possible.

If you must work on your rifle, be sure to unload it. If a round is jammed in the rifle, remember that it could go off at any moment. Stay clear of the barrel and keep the muzzle pointed in a safe direction. If a cartridge becomes jammed only part way into the chamber, treat it very gingerly; if it goes off it will pepper you with brass fragments. Always wear safety glasses and be very gentle when live rounds are involved.

Unless you are really sure of what you are doing, troubleshooting should not be attempted unless you are in combat and your life is on the line. Only then can the following measures be applied. Read them over and try them out with an unloaded gun. Do not attempt the actual measures unless you have no other choice.

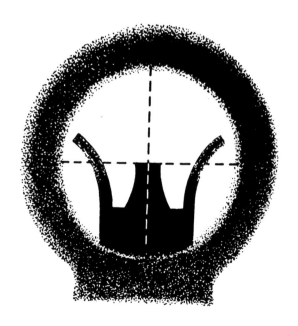

Correct sight alignment

Correct placement of the aiming point

PROBLEM	WHAT TO CHECK FOR	REPAIR PROCEDURE
1. AR-15 will not fire.	Selector on "safe"	Place in fire position.
	Wrong positioning of firing pin	Reassemble so that retaining pin is between the two shoulders of the firing pin.
	Too much oil or dirt in firing pin recess	Clean off oil and dirt.
	Poor ammo	Remove and discard.
	Broken firing pin	Remove and replace.
	Defective, worn or broken lower receiver parts	Remove, clean, and/or replace parts.
2. Bolt will not unlock.	Dirty or burred bolt	Clean or replace.
3. Rounds will not extract.	Broken extractor	Replace.
	Dirty/corroded ammo	Remove (may have to be carefully pushed out with cleaning rod).
	Carbon/fouling in chamber or extractor lip	Clean chamber and lip.
	Broken extractor or bad spring	Replace.
4. Rounds will not eject.	Broken ejector	Replace.
	Frozen ejector	Clean/lubricate.
	Bad spring	Replace.
5. Rounds will not feed.	Dirty or corroded ammo	Clean off ammo.
	Defective magazine	Replace magazine.
	Dirt in magazine	Clean and lubricate magazine.
	Too many rounds in magazine	Remove several rounds
	Poor buffer movement	Remove buffer/spring and clean and lubricate them.

PROBLEM	WHAT TO CHECK FOR	REPAIR PROCEDURE
	Unseated magazine	Magazine catch may need to be tightened. Push release button down and tighten or loosen the catch by turning it.
6. Double feeding	Defective magazine	Replace.
7. Round will not chamber.	Dirty or corroded ammo	Clean ammo.
	Damaged ammo	Replace.
	Fouling in chamber	Clean with chamber brush.
8. Bolt will not lock.	Fouling in locking lugs	Clean and lubricate lugs.
	Frozen extractor (in down position)	Remove and clean extractor.
	Buffer/spring do not move freely	Remove, clean, and lubricate.
	Bolt/bolt carrier do not move freely	Remove, clean, and lubricate.
	Bent gas tube	Check to be sure key goes over gas tube; if not, straighten tube as necessary.
	Fouling of gas tube inside	Replace gas tube.
	Loose or damaged key	Tighten or replace.
9. Short recoil (New rounds are not fed into the chamber.)	Gaps in bolt rings	Remove bolt and stagger rings.
	Fouling in carrier key or outside of gas tube	Clean key/tube end and lubricate.
	Missing or broken gas rings	Replace rings.
	Broken or loose gas tube	Replace or resecure.
10. Bolt does not lock open after last round.	Fouled bolt latch	Clean.
	Bad magazine	Discard magazine.

PROBLEM	WHAT TO CHECK FOR	REPAIR PROCEDURE
11. Selector lever binds.	Fouling/lack of lubrication	Lubricate; if it still binds, disassemble and clean.
12. Bolt carrier is hung up in the receiver.	Round jammed between bolt and charging handle	*Danger: Stay clear of the muzzle.* Remove the magazine; hold the charging handle back and slam the butt of the rifle against the ground. *Caution:* When round is freed, the bolt will remain under tension. While the bolt is held back, push charging handle forward and let the round fall through the magazine well.
13. Rifle will not cock; selector does not work properly.	Worn, broken, or missing parts	Check parts, replace.

15. Accessories

There is currently an almost-endless array of accessories for AR-15 rifles, and it appears new products will continue to be developed for some time. This makes it almost impossible to be familiar with all that is available for the AR-15.

Though a lot of accessories tend to be gadgets, there are some excellent devices available from time to time. Unfortunately, these often are not available for long, since the market for them is often small. Therefore, if you are really serious about getting worthwhile accessories for your AR-15, it is necessary to find out what is out there when it is still available.

SOURCES

There are four *good* sources of information, and a number of bad publications that purport to be good sources. The bad ones you probably already know: They cater to those most interested in "mean-looking" firearms and stories of blood, guts, and impossible feats. Such magazines never say anything bad about the worst devices.

The good sources are few and far between. Two I have found invaluable in writing this book are the *American Rifleman,* 1600 Rhode Island Ave., NW, Washington, DC 20036 (subscription cost: $15 per year); and *Soldier of Fortune,* Box 693, Boulder, CO 80306 ($26 per year; the articles by Peter G. Kokalis are especially good).

I do not place my own publication in the above group, but I do test a lot of equipment designed to be used with the AR-15 and review it in the *Long Survival Newsletter,* P.O. Box 163, Wamego, KS 66547. The price of a yearly subscription is $22.

Shotgun News (Box 669, Hastings, NE 68901) is a publication in which everything new being offered for the AR-15 by companies great and small can be found. It has *only* ads (no conflict of

interest there!). Whenever someone has something new in the way of firearms or equipment, it usually shows up in that publication first. *Shotgun News,* which resembles a newspaper's advertising section, has a yearly subscription rate of $15.

GENERAL ADVICE

Whatever you do, do not get overburdened with gadgets. It is easy to get caught up in the ad hype, as well as the pictures in combat magazines of young, burly guys with 120-pound packs. Nothing can get you killed more quickly in combat than a rifle that is impossible to use quickly and a load of gear that is hard to move in.

If you talk to guys who have fought for extended periods of time, you'll find the one thing they have in common is that they got rid of about half the junk they carried during training. In so doing, they can move quickly if necessary and can do so for longer periods of time.

Save some time and money, and do not get anything you can do without in the first place. Start with what you really need, and then stop. Unless you are a collector of military gadgets, you can get by with very little: your AR-15, a load of magazines filled with ammo, a good pocketknife, odds and ends of a first-aid kit, a canteen, suitable clothing, and maybe a butt-pack with a few necessities. Ideally, you will carry a lot less.

Be sure to test out all the modifications and accessories you have with your other combat gear. It is not conducive to survival to discover during a gunfight that your beautiful pistol harness makes it impossible to shoulder and aim your AR-15, or your rifle will not fire with your cherished night-vision scope.

I have tested almost, if not all, of the equipment in this book with the exception of some night-vision equipment. I do not receive any kickbacks from any of the manufacturers, nor do I sell anything to make a living (other than my newsletter and books like this one). If I have given a product good marks, it is because I think it is of good quality.

Usually, more than one company offers the products covered in these chapters, and I will only list the company that manufactures them or one or two good sources. If you are ordering a number of accessories, you may find it more convenient to order all from one company, even if some of the products are slightly more expensive.

One fault I do find with some otherwise excellent equipment is that they often have fasteners that need a hex driver or L-wrench. These drivers are easier to use than a screwdriver, but you do not want to discover that something is coming loose in the field and not have the right tool with you with which to tighten it.

For a while I carried the tools, but when you have them, it seems the need for them does not arise. Rather than burden myself at the cost of not carrying essentials (like candy bars), I put a slot in large screws and often replace hex nuts with wing nuts from the hardware stores.

To get the wing nuts, take the piece in and get a clerk to help you locate what you need. (A word of advice: Try to buy something in addition to the fourteen-cent nut so the clerk will wait on you again.) Blue the nut with cold blue. (See the section on building your own rifle.)

To slot a screw, use a triangle file to get the slot location started. With a hacksaw, slowly make a groove for a large screwdriver. Touch-up blue will make everything look like new again, except that now you have a screw that can be tightened in the field with a makeshift screwdriver (such as a chunk of metal or pocketknife). If you are careful, you can make a small enough groove to allow the L-wrench to be used, as well.

That being said, let us look at what is available. I will assume you have your ammunition and about six magazines for your AR-15. If you do not, by all means get them.

The stock, pistol grip, and handguards are the three areas where the AR-15 can be improved with commercial replacement parts. You might consider these changes, though you may find them unnecessary.

THE STOCK

Many tall shooters find the stock to be about an inch too short to be comfortable. A number of shooters have alleviated this problem by making a wooden insert which they place in the plastic stock after removing the trap door assembly. For a long time, this was the only solution to the problem.

Now there are at least two other routes. One is the new E2 stock from Choate Machine and Tool (Box 218, Bald Knob, AR 72010, 510/724-3138) for $30. This stock is three-quarters of an inch longer than the standard stock and seems just the right length for those folks who are taller than five feet five inches. At my suggestion, the stock also has a small hook so that the shooter can push it back against his shoulder when firing in the prone position. (While the prone position and bipod are not used that often in modern combat, it is nice to have the option, and it certainly does not detract from the stock's usefulness.)

The Choate stock is made of a new plastic, stronger than the old fiberglass stocks, and has more storage room in it. It is a quarter of a pound lighter than standard stocks. (Translation: You can carry six extra rounds of ammo.) Best of all, the swivel is a screw-in type (as found on a sporting bolt-action rifle) so it can be removed if it is not needed; if left on the rifle, it does not get caught in slash pockets like the standard AR-15 rifle swivel does.

The E2 stock is easy to mount. Remove the old

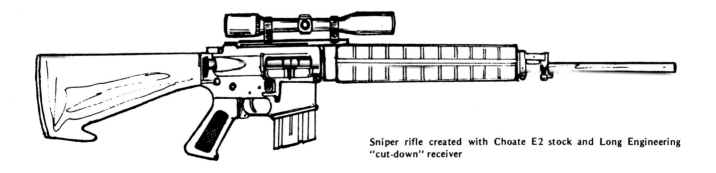

Sniper rifle created with Choate E2 stock and Long Engineering "cut-down" receiver

trap door assembly (by taking out the two screws), slip off the stock, slip on the E2 stock, put the trap door back on (a new longer screw comes with the stock), and it is ready to go!

The other solution to the problem of having too short a stock is to try to get some of the M16A2 stocks, which are also longer than the old stocks. These should be available on the surplus market, and I would be surprised if someone did not start making a commercial version of the longer stocks soon (or an extension insert to go into the standard stock).

PISTOL GRIP

The pistol grip on the AR-15 is fine as is, but a grip with a trap door in it allows you to carry a spare-parts kit, cartridge adapter (for firing .22s), small screwdriver, etc.

Years ago, GIs used an M1 Carbine magazine dust cover to cover the hollow grip. This rubber cap was *about* the size of the end of the grip so the cap could be stretched over the hole. Being *about* the right size, it usually did not quite fit. Because it often came off, it had to be taped in place.

The commercial trap-door pistol grip is a big improvement over the M1 Carbine arrangement. The trap door stays shut until you open it with a cartridge or pocketknife. The grip is also somewhat larger, so many of us with larger hands enjoy the bigger size.

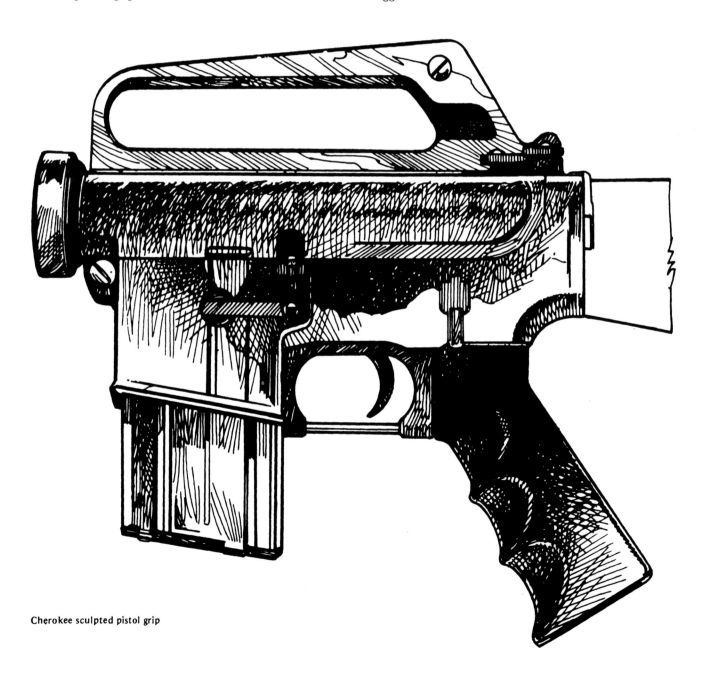

Cherokee sculpted pistol grip

Currently, just about every company that carries parts for the AR-15 is selling these grips. One good one is the Stow-Away Grip, available for $9.95 from Lone Star Ordnance (P.O. Box 29404, San Antonio, TX 78229, 512/681-9280).

Since the U.S. military is starting to use a new handguard on the M16A2, it is probable that the A2 grip and its look-alikes will be on the market as well. The new grip has finger ridges in it so that a steadier hold is possible. The best bet would probably be a pistol grip with both trap door and finger ridges; one is available on the commercial market. Although not the one actually on the M16A2, it is called the "M16A2 Contoured Grip."

The nylon "M16A2" is available for $15 from Cherokee Gun Accessories (830 Woodside Rd., Redwood City, CA 94061). It features three finger ridges and is three-eighths of an inch longer than the old-style AR-15 grip. The grip has a sliding cover at the bottom to create a storage compartment in the grip.

To mount either grip, use a large screwdriver to remove the screw and washer from the inside of the pistol grip, slide the new grip in place, and replace the screw and washer. (Occasionally the grips do not fit onto the lower receiver. If this is a problem, use a file to remove some of the plastic on the inside of the pistol grip.)

HANDGUARDS

The first production models of the AR-15 had rounded handguards like those of the 16-inch carbine barrel. These guards were on the first Colt AR-15, but were later changed to the "Beaver Tail" style. The round handguards were much more comfortable, making it much easier to keep a firm grip on the rifle in the rain or mud. Now, the "new" round handguards will be on the M16A2.

If you buy commercial round guards to replace the beaver tails on the AR-15, be sure that the upper and lower halves of the round guards are different so that they will fit into the triangular front retainer rim of the barrel.

Soon, surplus M16A2 handguards may come onto the market, and these *are not* interchangeable with the beaver tail stocks. The front retaining ring of the M16A2 is round, not triangular as found on the beaver-tail style. These round guards will not fit onto old rifles unless the front sight base is removed and a new round retainer is placed on the barrel.

Currently, round guards which are compatible with the beaver-tail style triangular guards are available from Lone Star Ordnance (address above) for $29.95 a set.

To replace the handguard, push back the weld ring (it may be necesarry to use a screwdriver to carefully lever it), lift the rear of each half and pull both halves out, slide in each new half, and let the weld ring spring forward to lock them in place. (If they do not seem to fit, it may be necessary to file a small amount of plastic off the rear, outside lip of each handguard half.)

ACCESSORIES FOR LEFTIES

If you are left-handed, there are some accessories you might consider necessities. One is an ambidextrous safety selector which places a second selector lever on the port side of the rifle as well as the bolt release side. This is available from L. L. Baston Co. (Box 1995, El Dorado, AR 71730, 800/643-1564) for $39.95.

The other accessory of use to lefties is a brass deflector. Some left-handed shooters are bothered by the brass that zings out of the ejection port. (Be sure your right arm is perpendicular, rather than diagonal, to the horizon. This will keep the brass from hitting your arm, while improving your accuracy.)

If you do not wish to be hit by brass, you may be able to locate one of the deflectors used for a time by the U.S. military. It was a simple, stamped metal piece that went over the top and rear side of the port. The deflector was held in place by spring tension and a bolt that fastened through the scope mount hole in the carrying handle of the AR-15.

The whole contraption is a little awkward, making it impossible to use a scope and difficult to carry the gun by the handle. However, it does work.

The new M16A2 model of the AR-15 has a bump that pushes the empties forward and away from lefties. Possibly a commercial rubber "bump" will come onto the market. If so, this might be something to look for if you are left-handed.

SLINGS

The standard AR-15, unfortunately, has sling swivels that seem to have been designed for marching rather than carrying the rifle. This often does not make much difference since the rifle is often used in combat with the sling removed. Sometimes, however, it is a bit aggravating. If you do use slings and may be in combat, try to place some type of quick detach device on each end of the sling so that it can be removed quickly if it is in your way.

The swivels often make a lot of noise when you do not have a sling on them. If you will not be using the swivels, remove the front swivel. Most new AR-15s generally have a stationary rear swivel. There is usually no need to remove this, though removing the screw that holds it, turning the swivel around so it faces the opposite direction, and replacing the screw can help keep the rear swivel from getting tangled in gear or pockets.

The AR-15 barrel is not supported by the handguard as are the barrels of many hunting rifles. This means that the AR-15 barrel flexes slightly if a large amount of pressure is placed on it. A problem arises if you try to use the sling to steady your aim. Apply a lot of pressure on the barrel with the sling while you are firing, and the impact of the bullets will drift to the side as much as three inches at one hundred yards. Slings are only for carrying the AR-15 rifle, not for trying to steady the weapon when firing.

Whenever you carry a rifle on a sling, put the safety on until you bring the rifle up to fire. A muzzle cap on the barrel is a good idea, too. These plastic caps are available from most surplus companies for less than fifty cents each and are well worth the buy. If you need your rifle in a hurry, you can fire through the cap and do not have to worry about whether the barrel is full of snow, mud, or dirt.

Most new AR-15s come with a sling that is just fine for marching or a walk in the forest. The tough nylon found in a lot of the slings will last a long time as well.

Probably at the head of the list for comfort and carrying ease are the numerous styles of slings available from Uncle Mike's (P.O. Box 13010, Portland, OR 97213, 503/255-6890) as well as most gun shops.

The Uncle Mike's slings run the gamut from the basket-weave leather Cobra Strap to inexpensive nylon slings which come in camouflage, brown, black, and white (for use in snow). The nylon slings come with padded "shoulder savers" or as regular straps. In general these slings cannot be beat (though I find their camouflage a little too light-colored for the best of hiding abilities).

Most folks in combat who want a sling prefer to carry the rifle in the "assault mode." This means that the sling is over their shoulder, and the rifle is ready for a quick shot from the hip or—better yet—

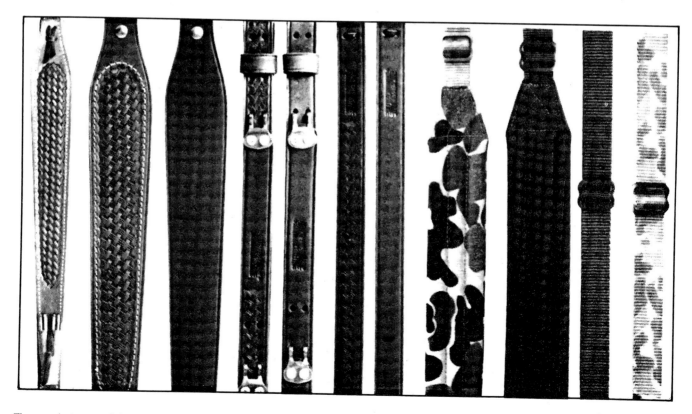

These are just some of the commercial slings for the AR-15 that are available from Michael's of Oregon.

waiting to be quickly brought to their shoulder. Besides the fact that the rifle can be quickly brought into play, many people find this method of carry easier than the old "marching" style of carrying rifles.

The assault carry is pretty easy to achieve with the commando-style telescoping stock: Thread the sling through the front sight mount (rather than the front swivel) and push the other end of the sling through the holes on either side of the buffer tube toward the rear of the stock. Adjust the sling so the rifle can be shouldered, and you are ready to go!

All is not quite so easy, however, with the standard AR-15. The problem is the AR-15's rear swivel location. There have been endless modifications and field expedient devices created to allow the assault carry with the AR-15, varying from threading the sling through the top of the front sight base to using shoe strings to fasten the sling to the handguards and top of the stock. All work to some degree, but none seem to work too well.

The Choate's E2 stock mentioned above has an extra three-quarter of an inch area at the upper rear of the shell of the stock, under which there is only one screw. If you make a plastic spacer the size of the buffer tube and put it between the end of the tube and trap door assembly, you can mount a swivel on the top of the stock by screwing it through the stock and into the spacer (needed for reinforcement so the swivel does not pull through the plastic shell). This top-mounted swivel requires a little longer sling for the assault carry, but it is a good solution to the problem.

Probably the best solution to the assault carry if you do not have the E2 stock is the Redi-Tac sling manufactured by JFS, Inc. (P.O. Box 12204, Salem, OR 97309, 503/581-3244). This system consists of two elements: a sling with a snap fastener in it and the counterpart of the fastener molded into a screw that replaces the top screw in the trap-door assembly that goes into the rear of the buffer tube.

One end of the sling is fastened to the front

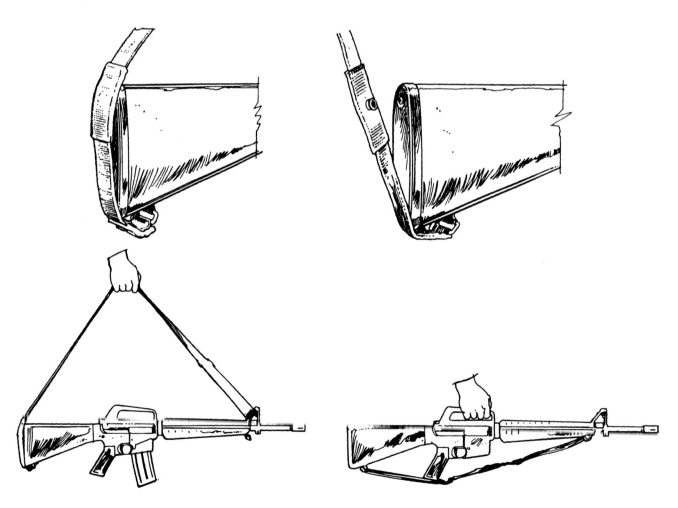

Redi-Tac assault sling system from J.F.S., Inc.

sight, while the other is fastened to the rear swivel. The snap is then put together so that the rifle can hang, right side up, from your shoulder in the assault position. When the rifle is needed, it can be quickly shouldered and fired since jerking the rifle up quickly unsnaps the sling and there is enough slack to place it on the shoulder.

All in all, the Redi-Tac is much more satisfactory than other assault sling carries. It will not get tangled in the charging handle when the rifle is cocked (unlike many "shoe string" or twine systems). The Redi-Tac will not work with an extended stock like Choate's E2 without a little modification, however, because the screw is too short to reach across the extra space (three-quarters of an inch) to the buffer tube. It is possible to alleviate this problem by purchasing the snap fastener counterpart to that in the sling (available at sewing stores or Five and Ten) and drilling out its center. Soft-solder (or silver-solder if you can) the snap onto the top of the screw on the Choate stock. Once on, it is possible to fasten the screw into the stock because the hole in the snap fastener allows you to use the L-wrench needed to fasten the screw in place.

It is undeniably a complicated procedure, but it is worth it if you need the E2 stock and want the Redi-Tac.

A kit that accomplishes the assault carry in a manner similar to many field-expedient systems is the "Top Sling Adapter" kit issued by the U.S. military. It is available on the surplus market from Sierra Supply (Box 1390, Durango, CO 81301, 303/259-1822) for $7.95. This kit consists of a spring wire piece that goes through the top of the front sight base and a loop that goes through the rear sling swivel and up over the top of the stock. Once these are mounted, a standard-length sling can be fastened to the two ends of the rifle and carried in a ready-to-go position. If you wear the strap over the shoulder on the opposite side from the rifle (i.e., over your left shoulder if you are right-handed), the strap should be adjusted so that it is easy to shoulder the rifle and quick aim is possible without your having to fight the sling.

Both ends of a standard-length sling can be connected to a bolt, wire, nylon cord, or other expedient device in the hole in the AR-15 carrying handle. Like the other systems, this allows the rifle to be carried over the left shoulder (if you are right-handed). If you need to use your hands, you can sling the rifle up over your back where it can

be brought around and into firing position in a hurry. (Keep a muzzle cap on the barrel so you do not get anything in the barrel while the rifle is slung behind you with its barrel pointing down.)

Six-foot nylon straps are available from a number of companies which carry AR-15 accessories. With a strap this long and a couple of extra slide buckles, it is possible to improvise a sling system similar to that used on Heckler & Koch combat rifles. This system has a lot of pluses, one being that the weapon can be worn very close to the body at a port-arms position and will stay there without the wearer having to hold it. This leaves the user's hands free for other tasks while the weapon remains ready for quick use; if the rifle is needed, it can be quickly swung up to the shoulder and fired since the sling adjusts with the movement.

To achieve an H & K "sling" arrangement, get a six-foot sling with a slide fastener on each end and an extra fastener threaded on so that the sling runs through one of the two spaces on it while the other hole is free. The stock must have a nylon cord run through the rear swivel and up over the top of the stock. With the regular AR-15 stock, tighten the cord so that it cannot slide over the back of the stock; with the telescoping stock, thread it through the slot under the buffer tube and up around the tube through the upper slot.

Fasten one end of the sling to the front sight assembly, and run the sling back along the selector side of the rifle with the empty half of the free-sliding fastener pointing outward.

Thread the sling through the cord tied around the stock, and bring the sling back toward the front sight. Fasten the free end of the sling to the open slot of the spare slide fastener.

The result is a sling that runs along the length of the rifle, doubles over under the cord on the stock, moves up along its first half, and ends on a slider which can move freely toward either the front sight or the stock.

To wear it, put the outside loop of the sling around your back and left shoulder (if you are right-handed), while the half of the sling next to the rifle should go across your chest. The rifle can now be slung so that it is either across your chest or pulled out and down in a firing position.

The AR-15 can also be slung back and up so that it is carried behind you with the barrel down (a muzzle cap is recommended for this so that mud, dirt, or snow do not get into the barrel

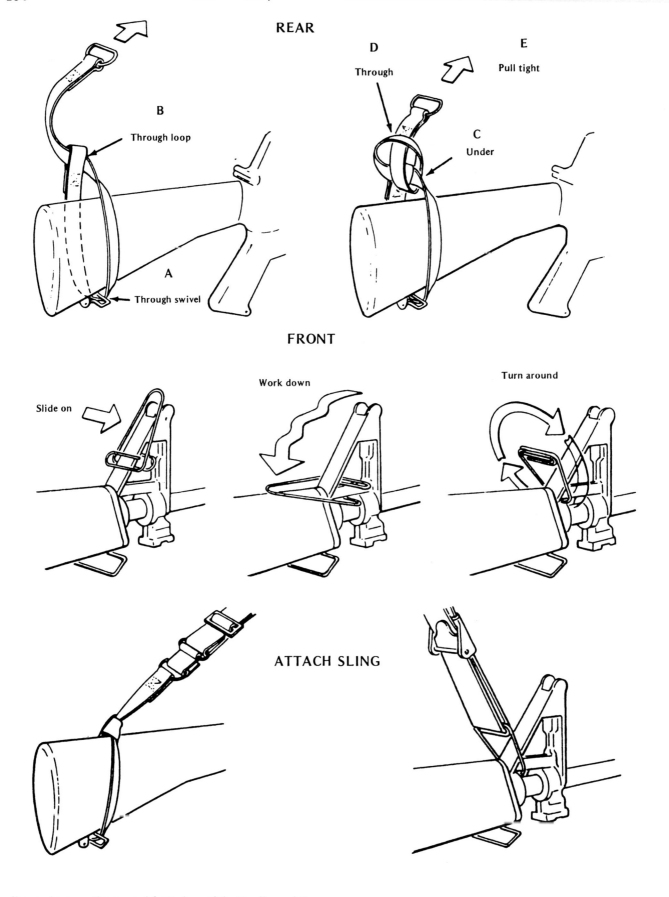

Shown above are the rear and front views of the top sling and the manner in which it is attached to the rifle.

should you kneel down). The rifle can also be taken off and carried with both straps together in the regular "marching" style. It can be worn on the back with an arm through each half of the sling, provided you do not require immediate access to the rifle.

All in all, the H & K sling is easy to create and is very convenient when you are moving around a lot and need access to your rifle.

BRASS CATCHERS

If you reload, you can save yourself a lot of groping in the grass and weeds if you get a brass catcher.

The old-style catcher was metal-framed and had a cloth bag which sometimes worked without jamming your rifle. A much better one is now available from E & L Manufacturing (2102 W. Coolbrook, Phoenix, AZ 85023) for $24.95. Called the Rigid Brass Catcher, it works well and can save you a lot of time and money. It can hold several magazines' worth of empty brass, rarely causes the AR-15 to malfunction, and snaps on and off, thereby making it easy to take off and empty. It is not, however, recommended for combat use.

you follow all the legal procedures, of course). Broken down, it is less likely to appear to be a rifle and less likely to be stolen than when it is in one piece.

A lot of people find it just as easy to carry their rifles around in an old cloth bag. I must confess that I have carried AR-15s, separated by releasing their push pins, in an army surplus laundry bag.

Carrying cases, however, can be much handier. Care must be taken in picking the case you need, since most problems arise when the wrong length is selected. The commando-telescoping stock style of AR-15 needs a 33-inch case; the standard rifle or a 16-inch barrel on a standard stock fits into the 41-inch case. If you have a 24-inch barrel, you need a 48-inch case designed for an FN FAL or one of the pistol-gripped 7.62mm rifles.

Metal cases with foam filling are available for those folks who travel a lot and leave their firearms at the mercy of airline loading crews. Since these cases demand a premium price and also tip off their contents to the casual observer, it is probably better to break down your rifle (as outlined above) and ship it in a box. Be sure to have a lot of padding around your rifle.

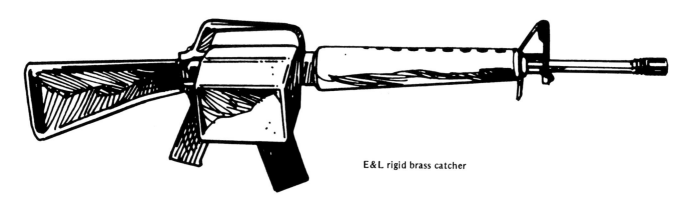

E&L rigid brass catcher

RIFLE CASES

Cases are good to have if you carry a rifle in a car or truck, store your rifle, or carry it through a populated area. The rifle case protects the rifle from small bumps and scratches, as well as the questioning eyes of people who might be alarmed by the sight of an AR-15.

If you store your rifle for a while and have a model with front and rear push pins, you can easily break the rifle down and store it in an old suitcase or box. This is probably a good way to ship it or to check it in as luggage when traveling by air (be sure

For storage or when carrying your rifle in a vehicle, the soft-shelled rifle cases are good and less expensive.

The Rolls-Royce of soft cases is the Assault Systems Cordura nylon case available from Assault Systems (869 Horan Dr., St. Louis, MO 63026, 314/343-3575). Available in black or camo, it has inch-thick padding to protect the rifle, and the nylon shells last almost forever. The case has five pockets for magazines (each of which will hold a 40-round magazine) and a long pocket for stowing a pistol, bayonet, and/or other odds and ends along with your rifle. In addition to the carrying handles,

Assault System's rifle case can hold a wealth of gear, as well as the AR-15.

it has a shoulder strap which can be detached for use as a rifle sling. A strong zipper keeps the rifle in place, and Velcro closures are used on the side pockets.

The Assault Systems case is a good choice if you will be carrying your rifle around a lot, and it is well worth the $65 price tag.

If you only need a case occasionally, it makes sense to get one of the less expensive cloth models. The best ones are made of a heavy canvas covering (camo, black, or olive green) with a lint-free, corduroy lining. They have a large, zippered pocket that will hold up to six 20-round magazines and a canvas handle that runs from the top around the bottom of the case so it cannot be torn off. They are available for $28 each from Parellex (1285 Mark St., Bensenville, IL 60106, 312/776-1150) or Sherwood International (18714 Parthenia St., Northridge, CA 91324, 213/349-7600).

BAYONETS

The original bayonets, which were spiked, were developed to replace the pike, a weapon used to defeat armored knights. By the mid-1800s, the man on horseback was becoming scarce, and the sword had become more of a decoration for officers than a tool of combat. Bayonets were then transformed from spiked blades to short, sword-bladed weapons. By World War I, most troops were armed with a long-bladed bayonet, which was seldom used due to the introduction of automatic weapons. Such weapons often mowed down bayonet charges before soldiers could even get to within shouting distance of one another.

When used, bayonets proved to be ineffective. The flat blades created too much friction when an enemy was impaled on them. Coupled with the extra leverage of a two-handed thrust, the friction often made it impossible to remove the blade from the victim's body. Troops soon discovered that the sword bayonet was about as useful as a sword, and were soon arming themselves with clubs, pistols, and trench knives.

In 1924, the British Army studied the problem and concluded the following:

- A long bayonet was not useful in fighting.
- A six-inch blade was the maximum length for body thrusts at an enemy in the heaviest of winter coats.
- Bayonets designed to be used as saws, machetes, etc., worked poorly as secondary tools.
- A long bayonet (if it did not catch in the enemy's body) often bent or broke, even in straw training dummies.
- Even when properly sharpened, the bayonet was often abandoned in battle once wedged in an enemy's body.

Rather than abandon the bayonet, the British issued a short spike bayonet that had no handle. The military users of the British spike bayonet, however, demanded a blade on the bayonet, and a short, bladed bayonet without a handle was issued during World War II.

In the United States, troops had carried a long bayonet during World War I and the beginning of World War II. However, troops were successfully using the M3 general-purpose knife rather than the bayonet when engaged in hand-to-hand combat.

To make the bayonet more useful, it was modified so that it had a blade similar to that on the M3 general-purpose knife. It could thus be used as a knife most of the time or mounted on a rifle to serve as a bayonet when needed for guarding prisoners.

During the early Fifties, a move was made to abandon the bayonet altogether and replace it with a good fighting knife. The first AR-15 models did not have a bayonet mount on them, but the M3 style of the bayonet lives on to the present day.

Brigadier General S. L. A. Marshal wrote in the May/June 1967 issue of *Infantry* that the bayonet was never used as a bayonet in Vietnam. Rather, troops in Vietnam learned that empty rifles, entrenching tools, and hunting knives did a better job in hand-to-hand combat where the battlefield was too overgrown with vegetation.

Bayonets, poor choices in combat, are effective for use in prisoner or crowd control. If you feel a need for a second weapon, get a small pistol or a good knife.

A bayonet may fill the bill if you need a good combat knife. Remember that the new United States bayonets have been designed after the M3 that proved so effective in combat. Many fighting knifes are expensive, but military surplus bayonets are not. Colt M7 bayonets (designed for the AR-15) are available from Sherwood International (address above) for very reasonable prices. If you are looking for a good camp knife for prying, cutting, or daily camp chores, the M7 bayonet has a rather poor blade for anything but finding mines or hand-to-hand fighting. You may wish to check out some of the Elkhorn bayonets which have a flat, single-edged blade (similar to those used by the USSR in parades). This flat blade can be purchased in a ten-inch model, which is excellent for cutting brush, and it looks impressive when you are mounting it on an AR-15 for parade purposes.

A good knife can be a useful replacement to the issue bayonets. One excellent knife designed for survival use is the Lifeknife. It is available from Lifeknife, Inc. (P.O. Box 771, Santa Monica, CA 90406) for $40. It has a hollow handle, which can be filled with fish hooks, wire, matches, etc., and has a compass and wire saw in its grip. The Lifeknife; various bayonets; and hunting, survival, and combat knives offer a wide selection of possibilities to anyone willing to do a little research to find the implement which best suits one's needs.

Never shoot an AR-15 with a mounted bayonet if accuracy is needed: the weight of the bayonet can change a bullet's point of impact when the barrel flexes from the extra weight.

The bayonet will continue to be almost useless on the end of a rifle unless the military designers go back to the spike bayonet. Knife bayonets will become increasingly ineffective as soft body armor and flak jackets become common on the battlefields. Strangely enough, the first bayonets came onto the scene because of hard body armor, and bayonets will probably become completely useless because of the new soft body armor.

SPARE PARTS KITS

It is wise to carry a few spare parts for your rifle. These can be placed in a pouch, but the best place is with the rifle. The hollow "Stow-A-Way" pistol grips are great for this, as is the hollow stock on the AR-15.

If you wish to make your own kit, it is easy to buy the parts from a mail-order company (check the section on building an AR-15 for names of companies). If you go that route, be sure to get a couple of firing pins, an extractor and its spring and pin, a cam pin, and a firing pin retaining pin. With a few additional odds and ends (a hammer spring, for example) your weapon can function for a long time if well cared for.

Kits containing most of these parts are also available from Sherwood International (address above) and Lone Star Ordnance (Box 29404, San Antonio, TX 78229, 512/681-9280). The parts in these kits actually cost less than when the parts are bought separately.

Acquiring two AR-15s—one for cannibalization for parts—does not work too well since (as with most rifles) some parts are more apt to break or wear out, while others remain in good shape. It is, of course, less cumbersome to carry a spare parts kit than two rifles!

When you order parts, be sure you get the right ones. If buying a trigger or sear, be sure to check whether you are getting the auto or semi version.

Even if you do not plan on repairing your rifle, having some parts available might make it possible

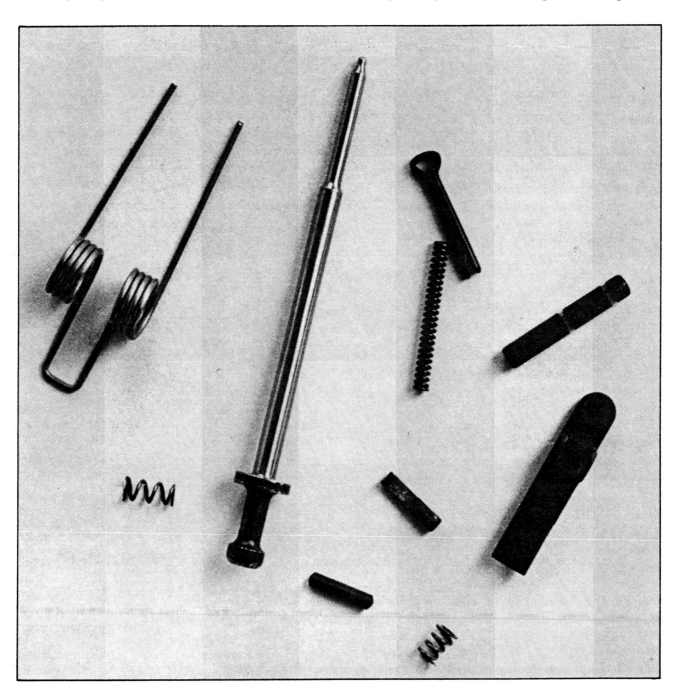

Spare parts kit

for someone else to work on your rifle. During a crisis, spare parts are very scarce, especially those for a rifle used by the military. Some other parts you might wish to acquire which are hard to make by hand are a gas tube, a hammer and its spring and pin, trigger and its spring, selector, and complete bolt. A complete barrel might also be a good idea if you can afford it. (Barrels, however, last for upward of 10,000 rounds if they do not get super hot on automatic fire or maximum loads.) Most other parts for the rifle can be made or improvised by a good gunsmith.

SILENCERS

Some may argue that the term "silencer" is inaccurate, but that was the name the inventor, Hiram Maxim, gave to his first silencers. If you want to be precise, you can call them "sound suppressors."

Probably more people ask me about silencers and automatic conversions for firearms than anything else. I almost always try to discourage them from obtaining a silencer—legal or otherwise—for the AR-15. Even with a silencer, the AR-15 is not silent with regular ammunition; with the ammunition to make it quiet, it is not too effective. One manufacturer whom I interviewed for this book may have said it best: "Silencers with regular 5.56mm ammunition are a joke."

Let us look at some facts.

An AR-15 with a 20-inch barrel creates a peak sound level of 165 decibels at one yard from the rifle (and a bit more with a shorter barrel). To give you some idea of how high this decibel level is, consider that an average conversation reaches 50 decibels; heavy traffic, 80 decibels; a thunder clap, 110 decibels; and the threshold of pain (where the ear starts to suffer damage) is 120 decibels. A .22 Long Rifle creates 135 decibels one yard from the gun.

Needless to say, it would take a very effective device to soak up the 165-decibel sound of an AR-15 muzzle blast. There are also other sounds which need to be quieted: the expanding gases hissing through the ejection port, the clattering of the action, and the supersonic crack of the bullet.

Despite what you may have seen on TV, there is no device that can make regular 5.56mm bullets quiet. About the best that technology can do is make the muzzle blast sound "only" as loud as a .22 rifle—which is far from quiet (especially compared to those Hollywood silencers!).

The cracking sound of the 5.56mm bullet is produced as the bullet breaks the sound barrier. When the bullet passes your area, you hear the sound, and it sounds as if it originates in the area you are in.

Silencers can be useful to a military sniper operating at long range. The muzzle sound of an AR-15 with a silencer will be "eaten up" by vegetation and terrain so that whomever the sniper is shooting at several hundred yards away will hear the crack of the bullet but not the muzzle blast. Forget about shooting at someone closer than two hundred yards without being noticed!

There is another route or two you can take, though. One route is to lower the speed of the bullet to below supersonic speeds. This is not a good option since the 5.56mm creates its damage by its high speed. Even so, the bullet still maintains more force at one hundred yards than a .22 Long Rifle does at the muzzle, so it is capable of doing some damage and would certainly be demoralizing. In addition to not having the crack of the standard supersonic rounds, the ammunition is also a bit quieter so that a silencer has less noise to mask.

If you reload, you can probably produce your own subsonic ammunition (a less-pointed nose on the bullet will also be more effective, but be sure it will still feed reliably). If you do not make your own subsonic ammunition, several companies manufacture it. One worth contacting is American Ballistics Co. (P.O. Box 1410, Marietta, GA 30061). With a good silencer and subsonic loads, you should be able to drop the sound down to the range of 100 to 120 decibels. The sound will still be far from silent, though.

It should also be noted that the subsonic loads will have to be fed into the chamber manually since the rounds lack the power needed to cycle the action. The weapon will be, for all practical purposes, a bolt-action rifle.

If you must get a silencer, get a good, legal one. They *do* work. Homemade ones must be registered *before* you start building them, and you must go through all the red tape needed to buy an automatic weapon. Keep in mind that homemade silencers often do not work very well.

When you buy a silencer, it should have serial numbers on the tube. You should also be prepared to pay the $200 transfer tax. Be sure to get all the paperwork done *before* you buy a silencer.

I hope you can see why I do not recommend silencers for an AR-15 to most people. If you want quiet shooting, use a .22 adapter or—if you do not mind cycling the action by hand for each shot—use a .22 adapter with CCI CB Caps. The CCI CB Caps are only good within fifty yards, but they are quiet and you do not have to spend $200 just to have the right to shoot them.

More detailed information about silencers is given in J. David Truby's excellent book, *Silencers in the 1980s,* from Paladin Press (P.O. Box 1307, Boulder, CO 80306, 800/824-7888) for $12. Truby gives an excellent overview of the types of silencers available, as well as the names and addresses of major silencer manufacturers.

BIPODS

A bipod is great for displaying a rifle. Often that is about the only use you will have for it, since most shooting in combat is done while sitting, kneeling, or standing due to ground vegetation. The vegetation makes it impossible to have a clear shot at an opponent when shooting from the prone position.

If you are firing from a prepared, fixed position with a clear field of fire or are using an AR-15 on automatic as a SAW, a bipod might then be useful. Occasionally, a bipod is useful for fighting in an urban environment where the rifle can be placed on rubble or ledges. Even so, a pile of rags often works just as well.

Adding insult to injury is the somewhat flexible barrel of the AR-15. With a bipod in place, the point of impact may change slightly on many rifles, especially if some of the shooting is done standing with the bipod on the barrel, while some is done with the bipod resting on the ground. Though the change of impact is not enough to matter in combat, it does not help things much. You may therefore decide that a bipod is not up at the top of your list of things to buy.

Should you decide to buy a bipod, the most common one is the metal "clothes pin" bipod used by the U.S. military. Made of either stamped metal or nylon, the bipod is spring-loaded so that it clamps onto the barrel just below the front sight base. The bipod is not very handy to carry, since the spring causes it to pop open all the time. It works, but that is about all that can be said for it.

A better bipod is made by B.M.F. Activator (3705 Broadway, Houston, TX 77017) for $12.95. It is made of a mixture of 66-percent nylon and 33-percent glass fibers that make the bipod quite strong. A big plus is the locking lug on the feet of the bipod which makes it possible to lock the bipod legs together when the bipod is carried.

The best available bipod is the Harris bipod, which can be purchased from Harris Engineering (Barlow, KY 42024) for $44.50. It works very well and is often found on military and police sniper rifles. Its legs are spring-loaded to lock them in either an open or folded position. Each leg is also capable of being opened to a different length for use on rugged terrain.

The bipod can be mounted on the AR-15 by removing the front swivel and enlarging the hole slightly. After this is done, the unit clamps into the hole and is tightened so that the bipod is firmly attached. A ring on the unit allows you to still use a sling on the rifle if you so desire. When it is needed, the Harris bipod can be quickly unfolded from the AR-15 on which it is mounted.

The Harris bipod comes in two styles: the 1A-LR, which has standard-length legs, and the 1A-HR, which has extra long legs that allow you to shoot from a sitting position—an especially useful feature in combat.

Ideally, the bipod would also be mounted from the area of the hand grip where you normally hold the rifle when firing from the standing position; this aids in alleviating the problem of barrel flex. To do this, however, you need some sort of clamp-on, do-it-yourself device.

Many shooters get along fine without a bipod by resting a hand on a stationary rest and placing the rifle against the hand. This allows for improved accuracy without the hassle of carrying around a bipod.

A rather strange product has also found its way onto the shooting market: a short rod which is fastened onto the user's belt and extends up to the handguard of his rifle. This allows for a steadier shot because of the support offered by the shooter's waist. Though this rod does aid in getting a less shaky aim and, unlike the low-slung bipod, is not hindered by the lack of visibility, the extra work of carrying a telescoping rod and the time needed to set it up will probably take more effort than most shooters might want to expend.

SCOPES AND MOUNTS

The AR-15 is very accurate, and a scope can be used to improve a shooter's abilities. This is not without its problems, however, since many hunting scopes are not as rugged as metal sights and are easy to whack into bits of junk. However, new sights are coming onto the market which are much more rugged and designed with paramilitary needs in mind. It is possible that scopes may gradually

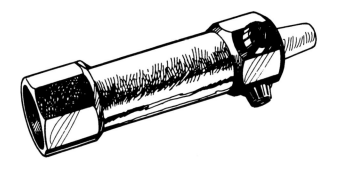

The Armson Occluded Eye Gunsight (left) and the Aimpoint dot scope (right)

replace the use of metal sights on combat rifles in the near future. In fact, sights can now be seen on several of the new bullpup rifles being developed by several countries.

Nevertheless, for sniper use or other long-range shooting, a good scope can make it possible to hit objects from unbelievable distances.

To be of use, a scope must be mounted securely on the AR-15. The carrying handle of the AR-15 was very well designed, having a long channel and mounting hole for attaching a scope mount.

One good mount is available from B-Square Company (Box 11281, Fort Worth, TX 76109). It comes with its own scope rings and is available only for one-inch scopes (the standard size in the United States).

While working on this book, I came across a mount that I must discourage everyone from buying. Made of plastic, it looks good, is inexpensive, and fits well. When a scope is mounted on it, the mount allows the scope to move when pressure is put on it. If you were using such a mount, it would be possible to get some dirt or twigs wedged under the mount and the scope would be sighted to some

spot far from where the rifle barrel was aiming. My advice is to stick with metal mounts.

If you always use your AR-15 with a scope, check into the Quicksight System described elsewhere in this book. It allows you to get a good cheek weld with a standard stock. The only drawback is that you will be without metal sights if the scope should become damaged. If you go with the Quicksight System mount, use a good, tough scope on it.

If you do not have a Quicksight System, you will discover that getting a good cheek weld is pretty hard. One solution is the Cheekpiece from Cherokee Gun Accessories (830 Woodside Rd., Redwood City, CA 94061) for $42. The urethane Cheekpiece attaches to the stock by a strap or Velcro connector. It allows you to have a proper cheek weld when using a scope and can therefore improve both the comfort and accuracy of your scope shooting. It is a little expensive, but it is worth the cost if you shoot a lot with a scope. (A do-it-yourselfer could possibly make a similar device for a lot less.)

One modification which can be made to the

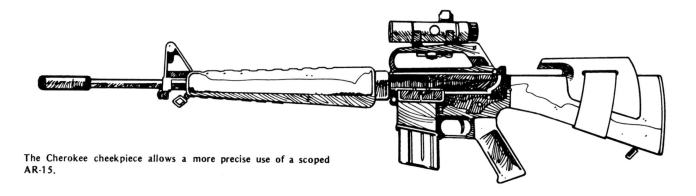

The Cherokee cheekpiece allows a more precise use of a scoped AR-15.

Cheekpiece is to cut off the appendage that extends out the front of it *provided* you do not need it. In heavy brush, the extension tends to get caught in branches and can be a real bother. It is there to allow you to still cock the rifle and have a place for your cheek. If you have an extended stock, chances are you can cut the extra projection off and never notice that it is gone. Just be sure to figure out where your cheek is while still allowing enough space to pull the charging handle all the way back.

Scopes are mounted onto the base or mount with scope rings. Because of the wide selection of rings, it is possible to greatly vary the height of the scope as well as use scopes of different diameters. If you use several different scopes on one mount, try to get rings that can be quickly and easily removed from the mount.

The best way to find the rings you need is to go into a gun shop with your rifle, mount, and scope and find some rings that work with your rifle and mount. Ordering rings through the mail can be frustrating when you do not know exactly how they will work until you finally get them.

No scope can make up for a lack of skill. If you are not much of a shot without a scope, you won't be much of a shot *with* a scope. On the other hand, if you are willing to spend some time at it, you can become a very good shot with a scope. Scopes are especially helpful, too, for folks that have vision problems that limit their abilities to use iron sights.

There are a number of choices to be made when getting a scope. One choice is the variable scope. It gives the user a choice of magnification, usually between three times and seven or nine times as large as the apparent size without the scope. These scopes are sensitive to abuse, and off-brand ones should be avoided because the zero may change with the magnification.

Range-finding scopes are usually variable scopes which can also move the zero up or down according to a range scale on the scope. These probably are not needed within three hundred yards. As we have seen, most combat shooting takes place within one hundred yards, so you probably can get along fine without a range-finding scope. Range-finding scopes are also usually slightly more sensitive to abuse, cost more, and have to be compatible with your ammunition. Think hard before getting a range-finding scope.

Most range-finding scopes work by having you bracket some object of known size between two cross wires. With hunting scopes, the known size

object is usually the animal's body; with combat scopes, a man's body. With a little figuring you can adapt the hunting scope to combat use or vice versa.

Once the object of known size is bracketed by the cross hairs, check the scale on the variable magnification control to see what the range is. (Some scopes have this range show in the scope picture, while others have it on the adjusting ring outside the scope.) Adjust the bullet drop compensator scale (being sure you have the right scale for your ammunition) so it reads the same distance as the scale on the magnification control. Adjust your magnification to suit your needs, and fire the shot.

Yes, it does take some time, though not as much as it does to explain the procedure!

Some very good scopes are the ARC and RAC by Burris, the BDC scopes by Bushnell, Redfield's Accu-Trac scopes, and Tasco's Range Finding-Trajectory scopes. Models are continually being refined and changed, so your best bet is to go to a gun shop and buy one of these companies' newest scopes which seems to best suit your needs.

The U.S. military used commercial scopes for snipers for a time but soon created their own ART (Automatic Ranging Telescope) system. Their system was quicker to adjust and capable of judging ranges. (Commercial scopes later followed the military's lead.) The first ART scopes were quite good and have led to a second generation of scopes. These ART II scopes have the range scale and magnification adjustment coupled together so that when the two cross hairs bracket an eighteen-inch object, the range compensation is right and the shot can be fired. This makes for very quick shooting.

The only bad thing about this system is that you are locked into a fixed amount of magnification in order to have the correct range. This is not as bad as it might seem, however, since greater magnification is needed for farther ranges in order to bracket the target.

ART II scopes are only for the serious shooter, since they carry a $600-plus-price tag. Unless you are really "into" long-range shooting, it might be a bit difficult to justify such an expense.

For a time, Leatherwood, the company which makes ART scopes, offered an MPC (Military/Police/Civilian) version of the ART scope. The scope was actually a Weaver variable scope as modified by the Leatherwood Company. Carrying a $350 price tag, it was more within the budget of most serious shooters. Unfortunately, at the time of this writing, the scopes no longer seem to be available. For information about the Leatherwood scopes, write to Leatherwood Enterprises, Box 111, Stephenville, TX 76401.

Fixed scopes are probably the safest bet for most shooters. They are tougher, less expensive, and last longer than the variable and range-finding scopes. The power most people seem to choose is the 4X scope. It gives enough magnification to be a help while still having a wide-enough field of view to allow you to quickly locate your target. Generally, the simple cross hairs are the easiest sighting device to use with the scopes if other methods are available.

Colt developed a small scope for the AR-15 which was aimed at the military market. The scope has a quick detach device and its own integral mount. One advantage this scope has is that it is small and therefore less apt to be bumped when the rifle is moved. It can be stored in a small space—even a pocket—when removed from the rifle. The Colt scopes are not cheap—$172 each— but the fixed, three- or four-power scopes are well made. They are available directly from Colt (Box 40,000, Hartford, CT 06151).

Several other companies have also introduced

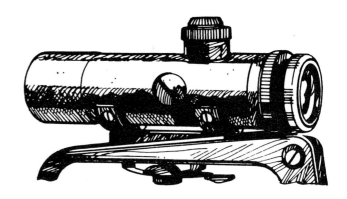

Colt AR-15 scope

small scopes modeled after the Colt; two of my favorite models are marketed by Beeman Precision Arms (47 Paul Dr., San Rafael, CA 94903).

The top line of the Beeman miniature scopes is the SS-2 which, despite its size, has a wide view and a coated lens. The lens is useful when the day is somewhat dark. This scope can be adjusted for parallax of the scope (normally not a concern, but a nice option to have). The height adjustment knob is made so that once it is zeroed in, it can be manually compensated for bullet drop on long-

Beeman SS-1 scope

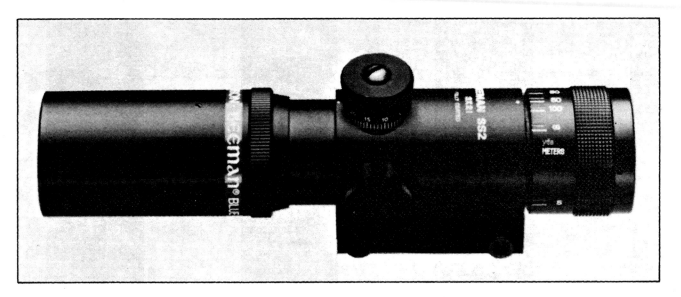

Beeman SS-2 scope

range shots. The SS-2 scope retails for $189.

Another version of the scope is the SS-1, which retails for $120. It is similar in size to the SS-2, but it lacks the easily adjustable bullet drop compensator and the parallax control. Because it lacks extra adjustment knobs (though it can be adjusted for windage and height) and has a tough rubber hide, the SS-1 is a much more rugged scope and will take a lot of abuse.

Both the SS-1 and SS-2 need special bases so that they can be mounted on the AR-15. These bases are also available from Beeman.

It is a good idea to keep up with the latest information as to what is available in the ever-changing scope market. Here are the names of some very good scope companies which will send you free catalogs if you drop them a card: Beeman (47 Paul Dr., San Rafael, CA 94903, 415/472-7121), Burris (Box 1747, Greeley, CO 80632), Bushnell (2828 East Foothill Blvd., Pasadena, CA 91107), Redfield (5800 East Jewell Ave., Denver, CO 80224), and Weaver (El Paso, TX 79915). Try to avoid off-name brands by companies other than these if you can. Unknown companies may have good scopes, but you never can be sure.

DOT SCOPES

In an age of electronics, it probably is not too surprising that there are a number of scopes with an electronic red dot rather than a cross hair. Most of these have only one power since their principle purpose is to speed up the sighting process.

When you look at a properly adjusted scope, a red dot appears at the point where the bullet will impact. If the scope is just one power, you can aim it with both eyes. This gives you a wide field of view and very quick aiming abilities. Consequently, these scopes have become the favorite of handgun contestants and are often used on AR-15 rifles when speed of target aquisition is critical.

Currently there are three brands of electronic scopes: Aimpoint (available from Parellex, 1285 Mark St., Bensenville, IL 60106) for $180, the Reflex Sighting System from Colt for $190, and the Tascorama Sight from Tasco for $239. These sights have a small battery that powers them for thousands of hours of use, and the brightness of the dot can be adjusted with a rheostat. Note that care must be taken in low-light situations so that the red dot's glare in your eye does not give away your presence.

All electronic sights are basically identical. The Aimpoint is probably the best buy and seems to be the choice of those engaging in shooting contests. (I am personally somewhat leery of something that needs batteries to operate. If you share that viewpoint, you might be interested in similar "dot" scopes that operate on available light.)

NONELECTRICAL DOT SCOPES

Nonelectrical dot scopes are only as bright as the light surrounding them, and both eyes have to be used to make them work.

To use the scopes, look at the target with both eyes and hold the rifle in the same normal aiming position you would assume when using a scope. The scope creates the illusion that a dot is superimposed on the spot where the bullet will impact. In fact, one eye sees the target; the eye which sees the scope sees only the dot. Your brain puts the two images together.

Like the electronic dot scopes, nonelectrical dot scopes make it possible for you to quickly find a target and fire upon it.

To get around the problem of the dot not being visible at night, most manufacturers add a small amount of a radioactive substance that creates its own available light by glowing inside the sight. It is thus possible to see a very light dot on your target when your eyes have become accustomed to the dark. The amount of radioactivity is about equal to that of a glow-in-the-dark wristwatch or smoke detector. The radioactive substance lasts for about ten years, at which time most manufacturers will replace it. If you are not too keen about having a radioactive sight, it is also possible to purchase the sights without the feature, which works fine in the daylight.

Care should be taken in the dark with the glowing sights, since they are visible within twenty or thirty yards. At night it is wise to cover the area in the front of the sight that gathers light so the glow cannot be seen from the muzzle end of the scope.

In my estimation the best available light scope on the commercial market is the Armson O.E.G. (Occluded Eye Gunsight). Along with its special AR-15 base, the Armson O.E.G. is sold by Armson (P.O. Box 2130, Farmington Hills, MI 48018) for $220.

A good runner-up to the O.E.G. is the Singlepoint. It costs less ($125) but the dot is bigger, which causes a slight loss in accuracy. The scope has a hump in the middle which requires the use of high scope rings if the scope is to clear the carrying handle when mounted on a regular scope mount (like that offered by Brigade Quartermasters mentioned above).

RANGE FINDERS

For the most accurate of shots when firing beyond several hundred yards without a range-finding scope or metallic sights, you must have some idea of the distance between you and the target. Bullet drop can make a big difference. Making a mistake in estimating the distance can mean a miss.

If you wish to accurately estimate distances, you need to have an instrument to verify your calculation. This is where range finders enter in. They can tell you the distances with amazing accuracy.

The best range finders available on the commercial market are being built by Ranging Inc. (90 Lincoln Road North, East Rochester, NY 14445, 716/385-1250).

The Ranging products are not expensive, and they *do* work. The company takes advantage of plastics to create a good product that would otherwise cost a small fortune if handmade lenses and tooled metal were used. Though not overly rugged, these range finders will stand up well and can be very useful with a little care.

Probably the range finder most useful to users of the AR-15 is the Ranging 1000. It gives distances out to 1,000 yards (you can order a metric model if you prefer).

The unit should first be zeroed in (a football field is handy for this), and you should practice with the finder at known ranges in order to become familiar with its use. The unit has a low-powered telescope which allows you to view images taken from the two ends of the range finder. Since the two views come from slightly different angles, it is possible to determine your distance from a target by the amount of adjustment needed to superimpose the two pictures.

Adjust the view until the two different images (which are easy to distinguish since they are of slightly different coloration) get together. Next, read the distance gauge. Provided you have done your job, the distance is right there for you to read.

The Ranging 1000 also has several range overlays that allow you to jot your bullet drop onto the scale.

The Ranging Company is working on a quick-check range finder which should be on the market in the near future. Though designed for hunters, it would also be useful in combat and for use in various combat games. Since most shots beyond three hundred to four hundred yards are apt to be misses, the range finder gives a color scale to quickly tell users at a glance how risky a shot is. For targets out to three hundred yards, the range finder gives the user the distance and a green "go"; for targets from 300 to 375 yards, the distance will be shown along with a yellow indicator; and for ranges over 375 yards, the user will see a red indicator. The real plus is that with an AR-15 zeroed on 250 yards, a user can probably get along without actually looking at the distance unless the range finder shows a yellow or red indicator.

The Ranging 1000 and the newer range finder are very useful tools. Write to the Ranging Company to find out about their complete line of useful ranging tools.

COMPENSATORS

With the AR-15, recoil is not much of a problem, but muzzle climb during automatic fire is. The muzzle quickly climbs to the right (for a right-handed shooter) and upward so that a lot of bullets are actually wasted.

A compensator ideally forces the rifle back into its former position after each shot. Though not necessary for semiauto fire, it is also ideal for semiauto shots since the compensator allows for a quicker recovery when a second shot is needed.

Compensators have been made which tackle the problem from both ends: the stock and the barrel.

Probably the most interesting compensator is the one created by John Kimball of Arm-Tec (23485 Industrial Park Dr., Farmington, MI 48024). This device was called the Fire Control Gun Stock and consisted of a hinged magnesium stock that flexed with each shot. The stock was then pushed into its original position by a spring and rubber buffer. This stock, to have retailed for around $100, enabled the shooter to fire a single twenty-round blast into a six-inch area at fifty yards. It was certainly quite an improvement considering that it was not necessary to wrestle the muzzle of the rifle to keep it on target (the stock did it for the shooter). There are some shooters who can achieve this kind of shooting without the stock—but only after many thousands of rounds and years of practice. Needless to say, such individuals are not too enthused about a device that makes a novice able to match their skills!

The stock did take a little getting used to since it was necessary that it not be held tightly against the shoulder when shooting—the opposite being true of a regular stock.

Though a number of countries tried out the rifle in military tests, to date none seems to be interested in the stock. It would seem the device was a little too late for the 7.62mm NATO and apparently not needed with the light recoil of the 5.56mm.

Another reason the stock may not have caught on was due to the inexpensive and effective muzzle stabilizers which were available at about the same time. Though there are a number of these on the market, the two best ones are the AK-74-style

from Alpha Armament (218 Main St., Milford, OH 45150) for $39, and the Mil/Brake, formerly called the Muzzle Mizer, from DTA (3333 Midway Dr., Suite 102-L, San Diego, CA 92110) for $25.

Both are threaded to replace the standard flash suppressor of an AR-15. Each of these compensators works, but each one creates slightly different effects.

The AK-74-style is modeled after the new compensator that was first seen on the Soviet AK-74 (their new .22 assault rifle). In addition to keeping the muzzle down, it slightly reduces the already meager recoil of the 5.56mm.

The trade-off with the AK-74-style is increased noise in firing, some problems with raising dust when shooting from a dusty building or the prone position, and a flash signature during nighttime use.

The Mil/Brake also works quite well. It has a flash suppressor which is nearly as good as that of a standard AR-15 so that flash is scarcely discernible during night shooting. The noise level of the Mil/Brake is also about the same as a standard AR-15 flash suppressor. The Mil/Brake does very well in compensating for a muzzle drift to the left or right, as well as upward, and can be adjusted to right- or left-hand shooting.

The trade-off made by the Mil/Brake is that it does not reduce the recoil of the rifle.

Both of these compensators can improve your recovery time when firing single shots, or greatly reduce the barrel displacement with full auto fire.

NIGHT SIGHTS

A very good night sight system was developed by the U.S. military and is often available to civilians via the surplus market. The system consists of a front sight which contains a small glass vial of radioactive promethium 147, and a modified rear sight with an extra large 7mm aperture for nighttime use.

The front sight screws into the front sight base just as a regular sight does. Sighting in with it is a little different with all but the M16A2-style sights since the front sight has only four detent positions. One "click" will move the bullet impact slightly higher or lower than one click of the standard AR-15 front sight. Care must be taken not to damage the somewhat fragile glass vial. The unit glows slightly and may be seen at close ranges in the dark.

During the day, the 2mm "L" sight is used

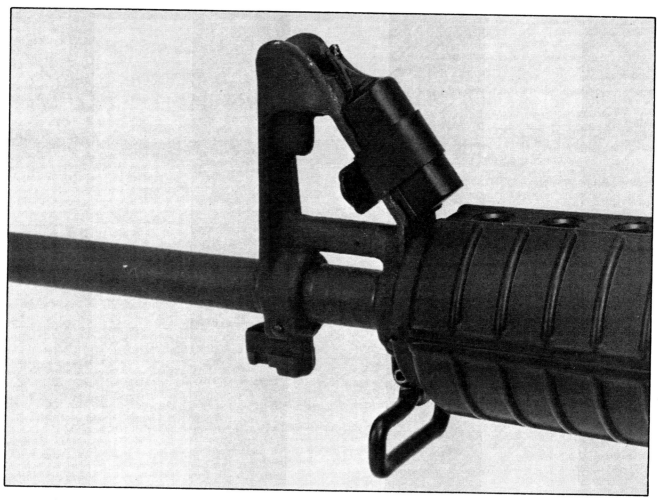

L.E.D. Nightsighter from Long Enterprises

and zeroed in at 250 yards. Beyond 300 yards it is necessary to practice hold-over rather than switching sights as is normally done with regular AR-15 sights.

With nighttime or limited visibility use, the sight is changed to the 7mm aperture. The front sight is centered in the rear sight and will have the same zero as the front sight.

It is wise to practice aiming the AR-15 during the daytime so that you learn to align the sights automatically. As a result you won't be firing above or to the side of the rear aperture when it cannot be seen.

Promethium 147 gradually loses its glowing strength over time and must be replaced. (The life of the sight depends on how old it was when you first purchased it; be careful you do not get an old surplus sight that hardly glows.)

Another device that produces the same effect as the promethium sight is the Nightsighter from Light Enterprises (P.O. Box 3811, Littleton, CO

80161, 303/770-9424). It uses a small light-emitting diode and two inexpensive digital watch batteries to create a tiny spot of light which can be lined up with the front sight.

This device is clamped onto the rear leg of the front sight and allows you to use standard AR-15 sights without changing them. It is thus easy to have the sights zeroed for regular or long-range firing during the night and then have the Nightsighter mounted for use in the same manner as the promethium sight.

The Nightsighter may look somewhat crude, but it works and holds up during poor weather. One source of contention is that the unit does not have a switch, so that it must be removed if you want to turn it off. Removing it can cause the batteries and lid to spill from the unit. On the other hand, the batteries will last for six months of continuous use.

The Nightsighter is worth considering with its low price tag of $18.

LASERS

The laser, unlike a regular beam of light, spreads out very slowly. While a beam from a flashlight may have a radius of several feet over a distance of several yards, the beam from the laser will be about the same size for ten, twenty, or even several hundred yards. It will only spread out slowly.

If a very intense beam comes from a laser mounted on a rifle, the beam can pinpoint the spot (within several inches) where an AR-15's bullet will impact over the first one hundred yards or so.

At the present time, there are two ways of designing a laser. One design creates a pinpoint beam, which will remain about the same size over several hundred yards (or even for miles!). The other creates a divergent beam or cone of light which will allow the laser beam to spread out about six inches over one hundred yards.

Each type of design has its advantages. The pinpoint beam can be seen even when it is relatively bright outside, while the divergent beam gets lost in bright light. Conversely, the divergent beam allows the shooter to aim toward its lower edge at long ranges so that bullet drop can be compensated for. The divergent beam tends to light up targets under extremely dark conditions so it is therefore possible to see the object you wish to hit.

You will have to decide which type best suits your needs.

The shooter, though, must keep the rifle rigid when firing; otherwise, a trigger jerk may make the gun point to a different spot when the bullet leaves the muzzle.

The spot of a laser is small and somewhat dim. It will not show what is being aimed at in total darkness (all you will see is the dot), and in bright sunlight the spot may disappear.

Nevertheless, laser aiming allows you to fire while the rifle is in any position as long as you can follow the bouncing dot.

Care should be taken when using a laser since the retina can sustain burns and a laser can leave a permanent blind spot. Treat the beam as you would a loaded rifle, and remember that a mirror or other reflective surface can cause the beam of light to rebound back to you. Protective glasses should be worn whenever a laser is being used.

Some lasers can be seen by someone down range from you, making you an easy target!

Other disadvantages of lasers include their cost and need for batteries. If you can afford the laser's price tag and don't mind being tied to a utility grid to recharge batteries, then the laser has a lot going for it.

Prices seem to be dorpping on laser-aiming systems, and the units are getting tougher and smaller. My advice is to look through *Soldier of Fortune* and *Shotgun News* to see what is available, or write to Laser Arms Corp. (P.O. Box 4647, Las Vegas, NV 89127, 702/648-2795) or Hydra Systems (Box 3461, Bridgeport, CT 06605) for more information.

NIGHT VISION SCOPES

There are two types of night vision devices: passive and active.

Active night vision devices date back to World War II. At that time, the United States used the infrared M3 Sniperscope, while the Germans had the Vampir sight. Active devices tend to be heavier than passive ones. They also have a smaller range since a source of infrared light to illuminate the area being viewed is required. Since the light and the optical system that turns the reflected infrared light into a visible image both need a lot of battery power, a heavy battery has to be lugged about to power them. Battery life is often quite short with this system. Worst of all, however, is that the light source is visible to other active or passive night-viewing equipment so that the hunter may quickly become the hunted.

On the other hand, active equipment is less expensive than the passive variety. Future use of divergent beam infrared lasers may extend the range of the units and limit the chance of detection by enemy troops with passive viewers.

Passive night-vision devices use available light: city glow, moonlight, starlight, or any other dim light. The advantage of these units is that their range is not as limited as that of the active night vision devices, and they are also hard to detect. Despite TV ads that would have you believe that passive devices were created to help those with night blindness to see again, the equipment was used during the Vietnam era, with the purpose of enabling U.S. snipers to kill enemy soldiers.

Passive devices are based around an image intensifier which is much like a TV camera. It takes small contrasts in the light and dark patterns of whatever it is pointed at and boosts the lighted portions (to perhaps 64,000 times the brightness of the original picture). The pattern is created on a screen, which is then viewed by the user.

The passive system must have some light in order to work. On an extremely cloudy night, the scopes will not allow the user to see *anything*. Such occasions are rare, but they sometimes do occur.

Currently, two generations of passive night-vision devices exist. The first generation is good, provided it is not necessary to view areas which are partly lit. In such a case, streaking and blooming can result. Streaking is caused when a light is viewed; the light causes a streak of light to form on the viewing screen for anywhere from a fraction of a second to a full second. Blooming is caused when a lighted area is viewed; it is a flaring effect that takes place around the object being viewed.

Though first-generation passive equipment suffers from these negative effects, it is cheaper than second-generation material, and the ability to view areas with lights is not always needed.

Second-generation equipment is state of the art, and is quite good. Mounted on a rifle, it can work both in the daytime (with a special filter) and during the nighttime. Night-vision goggles enable the user to see in the dark: You can use regular high-quality binoculars or rifle scopes and see through them almost as if it were daytime. They do, however, give off a green glow in the viewing area.

A good-quality passive scope for an AR-15 rifle is currently going to cost somewhere in the neigh-

borhood of $2,000 to $15,000. The bulkier (and currently much inferior) active equipment will range from $500 to $1,000.

New thermal imaging units are beginning to show up on the market. Though they currently are not suitable for use on an AR-15 rifle, it will probably only be a matter of time before they can be mounted on a rifle.

The thermal or infrared band is actually the same as radiated heat. Thus, anything that gives off heat can be seen with a thermal imaging system even if it is not visible to the eyes or is concealed behind a flimsy cover. The thermal imager could, for example, be used to inspect what may look like a peaceful, wooded area. Whether by day or night, it would be possible to see if it were a hiding place for soldiers ten yards inside the edge of the forest.

The only drawback to the above-mentioned devices is their cost. If you invest the large amount of money required to purchase such a device, be sure you get what you need and then take good care of it.

Passive units can quickly be put out of action when the lens is exposed to a bright light with the unit on. Always mount the protective covers during the daytime—even if the unit is off.

Like laser-aiming systems, the price of night-vision devices is currently dropping, while the units are becoming more rugged and reliable.

Probably the best route to go is to contact a number of companies, choosing units which best suit your needs. Be sure to ask for references, and

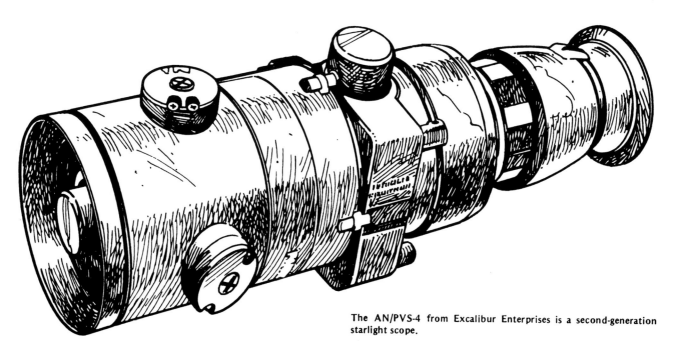

The AN/PVS-4 from Excalibur Enterprises is a second-generation starlight scope.

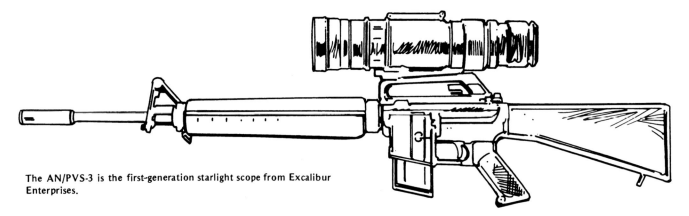

The AN/PVS-3 is the first-generation starlight scope from Excalibur Enterprises.

be prepared to offer references yourself since many state and federal officials are trying to outlaw night-vision devices, and companies are therefore trying to keep the units out of the hands of criminals.

For more information regarding night-vision products, contact Excalibur Enterprises (P.O. Box 266, Emmaus, PA 18049, 215/967-3443), Litton Electron Tube Div. (1215 S. 52nd St., Tempe, AZ 85281, 602/968-4471), and Standard Equipment Co. (9240 N. 107th St., Milwaukee, WI 53224, 414/355-9730).

CAR-15 VEHICLE MOUNT

The CAR-15 Vehicle Mount is designed only for holding a Commando-style carbine with telescoping stock. It is not designed for the standard AR-15. The mount was actually a government issue kit from the Seventies. In addition for use in a vehicle, it can also be mounted in a closet or near a door.

The finish on the CAR-15 Vehicle Mount kit is not attractive. The base is painted, and its inside and barrel clip are coated in a greenish vinyl plastic in order to protect the rifle.

The base cradles the stock of the carbine and the barrel clip locks the rifle in place, waiting to be jerked free. Both are mounted with screws or bolts to the position you desire. All in all, the CAR-15 Vehicle Mount can be very useful, and is available from William J. Ricca (P.O. Box 802, Clark, NJ 07066) for $17.

CAMOUFLAGE COVERS

Camouflage covers for the AR-15 are available from Military Surplus Supply (5594 Airways, Memphis, TN 38116) for $19. The sets consist of two cloth covers—one for the handguard and one for the stock—which are held on the rifle with Velcro fasteners. Since the covers quickly snap on and off, these fasteners are ideal should you need to occasionally camouflage your rifle. (Care should be taken not to allow the rear cover to get tangled in the charging handle.) Remember, too, that the handguard ventilation holes are blocked by the front cover; the barrel can overheat more quickly than usual with the cover on the rifle.

If you wish to have your rifle fully camouflaged, either paint or tape the rifle. Though paint scrapes off easily on the AR-15, it is often more satisfactory and allows you to create a camo pattern and color that more closely resemble your surroundings.

Tape is pretty easy to use. There is no need to place oil under it except on the barrel of the AR-15 since rust cannot develop on the plastic or aluminum. Except for the barrel, the tape can be left on the rifle until it wears off or you decide to remove it.

Camo tape is generally available at sporting goods stores or gun shops.

BLANK-FIRING DEVICES

Although most training on the AR-15 can be better done with subcaliber conversion kits, there occasionally may be a need for a blank adapter, especially for war games or reenactment groups.

Since a blank does not develop as much recoil as a normal round does (which is why Hollywood heroes have such great control over automatic weapons held in the most casual ways), an AR-15 will only fire single shots if a blank adapter is not used. These adapters plug up most of the barrel and force gas through the gas tube rather than allowing all of it to exit the muzzle of the rifle.

There are two types of blank adapters currently available for the AR-15. One is the "Hollywood type" which rides inside the flash suppressor of the rifle and is all but unseen from the outside of the

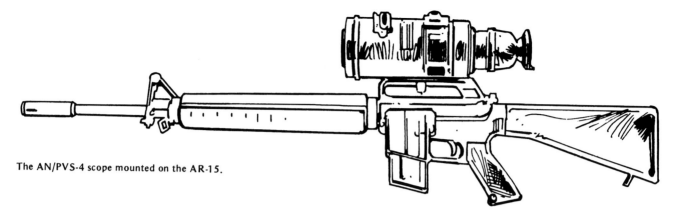

The AN/PVS-4 scope mounted on the AR-15.

rifle. This type, though giving a realistic appearance, can be rather dangerous since one can easily leave it in the rifle when switching to regular ammunition—with disastrous results! It is hard to determine whether firing a rifle with blanks is really safe for your human "target."

A better blank adapter is the type available on the surplus market. These fasten to the flash suppressor and have a bolt which screws into the muzzle of the suppressor. These adapters are generally painted red so that they are easy to see when they are in place on the rifle.

Though most blanks are loaded with smokeless powder, occasionally blanks are sold which contain black powder. For those fortunate enough never to have worked with black powder, the stuff is hygro-scopic and promotes rust and corrosion in any firearm in which it is used. Though the chrome-lined barrel of most AR-15s is not harmed by black powder, the same is not true of the gas tube, bolt, bolt carrier, or inner mechanism.

As previously mentioned, new accessories and modifications for the AR-15 are continually being developed. Do not get caught up in trying to get the perfect rifle and accessories—you never will. A person who is familiar with his equipment and can operate it well can almost always outperform the guy with state-of-the-art equipment who is bogged down by too much and hampered by ignorance.

Get only the essentials and practice, practice, practice.

16. The AR-15 in Combat

There is, of course, a very big difference between hitting a target and hitting an enemy in combat. Often there is no time to take careful aim at the fleeting targets encountered in combat. One way of countering this is to fire a number of bullets in the enemy's general direction. This approach works, especially at close range, but necessitates large amounts of ammunition to make up for the increased waste involved. While such a waste may seem justified—especially to the guy in the thick of the fighting—there are other ways to wage combat.

While overwhelming firepower seems to be the name of the game for modern armies and the automatic rifle is seen in the hands of most modern troopers worldwide, this situation could change. Many military studies actually suggest that three-round bursts are more effective than endless automatic fire and that there are limits to superior firepower. Spraying huge areas can never substitute completely for aimed fire. Some of the new technologies may create superior systems which use laser or glass-fiber optical aiming mechanisms. Such systems could conceivably change the use of the assault rifle.

Even when automatic fire is needed, it is not needed for any length of time, which is fortunate since AR-15 barrels get too hot if fired for long—especially since they fire from a closed bolt.

Suppressive fire can be achieved with forty to sixty rounds per minute. The AR-15 can handle this for a short period of time. To really pin an enemy down, though, a rate of one hundred rounds or higher is needed, and the need for fire might go up to two hundred rounds per minute during an ambush. There is no way the AR-15 can sustain such fire for more than a few minutes without destroying the barrel and creating cook-

offs. Obviously, if you must choose between your life or the destruction of your rifle, the choice is not hard to make. But remember that you only live to fight another day *if* you have a working weapon. Unless you are fighting your last battle or have a Squad Automatic Weapon with you, automatic fire is not as practical as one might think if needed for any length of time.

In a battlefield in which chemical, biological, and/or nuclear weapons are being used, supply lines may quickly become disrupted. Soldiers on modern battlefields may virtually be on their own, having only the ammunition they can carry (or steal) to get them through long weeks of battle.

Aimed fire and knowing how to pick targets can quickly become of prime importance to your combat survival even if you have an automatic AR-15. Having a little know-how and doing a little planning ahead can quickly make up for a lack of firepower in many situations.

One important point to consider is what a certain bullet will do in your AR-15. While some users of AR-15s have to make do with issued ammunition, shooters who can reload ammunition can create rounds capable of tumbling on impact or extensive penetration.

The following table should give some insight as to how to accomplish extensive penetration and/or tumbling. Military experts generally agree the ideal gyroscopic stability in combat use is 1.5 to 2.0. This range provides accuracy over a wide range of temperature and altitude conditions while still creating a round capable of tumbling within a target. Bullets with smaller gyroscopic stability ranges are very deadly but may become inaccurate during cold weather. Bullets with higher gyroscopic stability will penetrate barriers and have better long-range accuracy. With any given twist rate, the heavier the bullet fired from the barrel, the less stable it will be. Conversely, lighter bullets will have a higher gyroscopic stability.

The original AR-15 with its "tumbling" bullets fired a projectile with a gyroscopic stability of 1.0, while the M14 rifle bullet had a gyroscopic stability of 2.2 and the M1 Garand, 3.0. As the range becomes greater, the gyroscopic stability of a round increases because the bullet slows down in its forward motion more quickly than its spin slows down.

GYROSCOPIC STABILITY

55-Grain Full Metal Jacket Boat-Tail (M193), 3200 fps

1-to-14 in. twist: $S = 1.0$
1-to-12 in. twist: $S = 1.4$
1-to-9 in. twist: $S = 2.4$
1-to-7 in. twist: $S = 4.1$

62-Grain Full Metal Jacket Boat-Tail (SS109/M855), 3050 fps

1-to-14 in. twist: $S = 0.62$
1-to-12 in. twist: $S = 0.85$
1-to-9 in. twist: $S = 1.51$
1-to-7 in. twist: $S = 2.5$

Since 52-, 53-, 55-, 62-, and 69-grain bullets are available in 5.56mm on the commercial market to a reloader, a little study will show you how to create the round you may need.

Let us suppose you wish to use your rifle in a house-to-house fighting situation where you wish to obtain quick, lethal hits in close-range confrontations. In such a situation you will probably be more interested in having a short-range tumbling bullet rather than a superaccurate, long-range bullet, which might penetrate through walls (and injure innocent comrades or bystanders).

Let us assume you have a 1-in-12 twist in your AR-15. You can create a round with very poor stability by using a 62-grain bullet rather than the standard 55-grain projectile. If you had a new M16A2 with a 1-in-7 twist, you could improve the tumbling tendency of a round by loading it with the heaviest bullet available: the 69-grain bullet.

On the flip side, if you need a long-range sniper round or may have to fire through a car or window, you need a bullet capable of great accuracy and penetration. In such a case, you would need to go with lighter bullets or a faster rifle twist.

Those not barred by the conventions of warfare which prohibit soft-or hollow-point ammunition can have the best of both worlds when it comes to accuracy and lethality of 5.56mm bullets. By using a bullet with a high gyroscopic stability but with an expanding configuration, it is possible to create even greater wound effects than a tumbling 55-grain bullet creates. It is thus possible to have high accuracy while maintaining high lethality. (This is hardly a new idea; many troops in World War I and World War II filed the tips off full metal jacket bullets to create expedient expanding bullets.)

When using hollow-point bullets, great care should be taken to extensively test them in your AR-15. If the round does not feed well, as may sometimes be the case, you may not be firing anything, rather than firing the ultimate bullet. Soft-point bullets generally feed well, but again, it is wise to try them out beforehand.

Conventional bullets can also be altered to suit special needs.

BULLET ALTERATIONS

Suppose you need to use a bullet indoors. Normally the 5.56mm is a very poor choice if there may be innocent bystanders elsewhere in the house. Even though the bullet may tumble or expand (with soft or hollow points), it has enough energy to go through most modern houses or apartments while still having enough energy to be lethal. In such a situation, a full-metal jacket can be altered to make it expand more quickly so that it will expend its energy either in an enemy or harmlessly in Sheetrock and wood.

Most alterations consist of removing part of the metal jacket with a file. One good way to do this is to take a file and remove the outer metal jacket about a fifth of the way down from the tip of the bullet, leaving the lead tip exposed. Use a triangular file to create two cuts (about a fifth of an inch deep), which run from the bullet's tip and down its sides. When done properly, an "X" can be seen when you look down at the point of the bullet.

This method can be used to create ammunition for indoor fighting from regular military ammunition. (Try out the rounds in building materials like those you will be shooting in to be sure you have made the cuts deep enough. Finding out that you've killed an innocent bystander after a gun battle is not the way to discover that the bullets weren't altered correctly.)

If you reload, bullets designed for other .22

centerfire rounds (just so the bullet diameter is .224 inches) can be used. If you go this route, 40- and 45-grain bullets are available. These are altered in the manner outlined above. With the filed X and exposed point, they open up even more quickly than the altered regular 5.56mm bullets. Again, these should be tested out to see whether they open up and to be sure the recoil created by firing the lighter bullet will still cycle the action.

The 5.56mm—especially with the faster twists—is able to penetrate most soft armor vests as well as car bodies. Armor-piercing bullets are rarely needed with the AR-15.

Armor-piercing bullets can be created from commercial rounds. The procedure is time-consuming but simple. First file off the bullet tip so the lead in the point is exposed, and use a small drill bit to drill a small hole into the bullet. Hold the bullet in a pair of pliers while you screw a small screw (made of stainless steel or steel, not aluminum or brass) into the hole. File the exposed end of the screw down to a point. With a fast twist, this round will have very good penetration, so be sure of your backstop when test-firing this round. In combat, the bullet will have very good penetration of barriers and vehicles.

In addition to military and do-it-yourself armor-piercing rounds, it is currently possible to buy armor-piercing bullets from American Ballistics Co. (Box 1410, Marietta, GA 30061, 404/434-8078) for about $18 per box of twenty-five.

Before you go to all the work of creating armor-piercing rounds, be sure you need them. Though the 1-in-14 twist barrels send out bullets that may or may not penetrate the sides of a car, the 1-in-12 twist probably will (though it might have a little trouble with the safety glass). The 1-in-9 and 1-in-7 twists will zip bullets into nonarmored vehicles.

Unless you are going to take on an APC (Armored Personnel Carrier), you probably won't be needing the special rounds.

If you reload, use only brass that has been fully resized, and use only the best of components. If you only resize the neck, the round may fail to chamber when the rifle becomes dirty. Cheap components may make rounds unreliable.

Test-fire a few of your reloaded rounds to be sure you have rounds that will cycle the action. (Light loads often cause the rifle to fail to feed another round off the magazine and into the chamber.)

POWDERS

Probably one of the best powders for the AR-15 is the IMR 4198 series of powders. These vary slightly from one manufacturer to another, so be sure to check their charts before reloading with it. For accuracy, Hodgdon H335 works well, though bullets fired with it will often have a slightly lower velocity and the powder adds a little more carbon to the bolt of the AR-15. The more common IMR 4895 works, but it is inferior to both the IMR 4198 and H355.

In general, the faster burning rifle powders are preferable to slower powders since they will cause less fouling of the AR-15's bolt.

It is possible to make your own rifle and primer powder and to reload primers and cartridges, but it takes a lot of work and the powder may not be of good quality. The methods of making smokeless powder are quite dangerous, though not as dangerous as black powder (which would perform very poorly compared to other powders you can make). The procedures are outlined in my book, *Survival Reloading* (available from Long Publications, P.O. Box 163, Wamego, KS 66547, for $5.98).

MAKE-YOUR-OWN BULLETS

It is also possible to make your own bullets. Lead bullets are the easiest to make but cannot be fired at the high speeds normally used in the AR-15. To make jacketed bullets suitable for the AR-15, you need good swaging dies and some other odds and ends. A standard reloading press can be used to make the small 5.56mm bullet, unlike most larger rifle bullets which need a special press for swaging.

The problem with swaging is that you do not save much money and it requires a lot of time. Even if you had a free source of lead and used empty .22 Long Rifle cases to make bullets (which is possible), you would still need to make several thousand bullets at today's commercial prices to pay for the equipment—assuming you used your own reloading press and your time is worth nothing!

Swaging does have a big plus, though: You can make exactly the type of bullet you want—45 to 65 grains in weight, FMJ or soft-point, special bullets loaded with buckshot or steel penetrators, etc. The possibilities are limited only by your imagination.

For the ultimate in combat bullets, you need to swage them yourself. If you are interested in investing some time and money, you should write to Corbin Manufacturing (P.O. Box 758, Phoenix, OR 97535) for a catalog and information on their swaging equipment. Their products are probably the best—and cheapest—available.

GELATIN

Whether you use issued ammunition, store-bought, or build your own from scratch, sooner or later you will probably want to test-fire a rifle to determine how effective it might be in combat. Some materials, though, give very little indication of how a bullet would react in human flesh. Among the worst are duxseal, clay, and grease. (The gun magazines use these substances to compare rounds and create interesting photos.)

Blocks of ballistic gelatin (USP-A), which contain about the same amount of water as human tissue, are good targets for testing your rifle. The material is, however, expensive, does not keep well, and you would need a huge refrigerator in which to keep it if you plan to make large quantities.

THE "WET PAPER" TEST

All is not lost, however. The technical staff of the National Rifle Association recently determined that water causes bullets to expand and slow down at about the same rate as gelatin (or flesh and blood). So, too, would newsprint *if* it were saturated by water for at least eight hours and the shot was fired at the paper when the water was still running out of it.

Ballistic gelatin or soaked papers will show the minimum damage of the round since secondary missiles will not be created by metal gear or bones. If a round performs well with these types of tests, it will probably cause greater damage in living targets.

If you try the "wet paper" test, you will probably find something like this with a 55-grain solid point fired from one hundred yards: The outward appearance of the impact will not be spectacular (just a .22-caliber hole). As you remove the layers of paper, you will see that several inches down the bullet started tumbling and created a keyhole-shaped channel. By the time the bullet has traveled four inches, it has torn out a hole an inch or more in diameter which will continue for some distance.

Soft-point bullets create even more devastating "wounds": At two inches into the paper, the wound channel will be an inch in diameter. At three inches, it will probably spread out to an inch and a half, and at four inches it will be a two-inch hole with distortion and pulping within a four-inch diameter area.

DETERMINING AN ENEMY'S DISTANCE

Before you decide what bullets you will need, consider what combat conditions you may be in and look at the ranges from which you may be firing.

Let us look at a few basics that are often overlooked by those who prefer to substitute superior firepower for other basic tactics.

You first have to locate an enemy before you can engage him. In combat, a trained enemy will rarely reveal himself, except perhaps during an all-out assault. When you do see or detect him, he will usually be within three hundred yards from you.

Within this range, you will probably locate the enemy because you have seen some indication of his presence: a muzzle flash, sounds, or reflected light from some shiny object he is using. Rarely will you actually see your enemy in combat. If you do, it will probably only be for a brief moment.

You cannot locate your enemy if you are in a position which offers bad visibility. Try to get into a place with maximum visibility, while also finding a place that offers full cover or concealment.

If you move into an area, try to find the "ideal" position and get into it quickly. While doing so, search the area visually to be sure you are not in immediate danger from enemy observation. If you are already in an area which you will be defending, then you should have given a little thought as to where an enemy may head, what spots he might choose to hide in, etc. When you suspect an enemy is in an area, search the area visually just as you would if you were to move into a new position.

Your rapid search should be made during the first thirty seconds in a given area or when you first suspect an enemy's presence. Start looking at things close to you, gradually working your way out to three hundred yards.

You will miss seeing an enemy if you just sweep your eyes over the area. Focus instead on specific spots. When your eyes look directly at a number of separate spots rather than sweeping over the area, you are more apt to see someone who is around you.

If the enemy has not been sighted during the thirty-second search, go over the area visually in a slower search. This time, start with the fifty yards closest to you, gradually working your way from the far right to the far left.

If the enemy still has not given himself away, start with the next fifty yards and gradually work your way out to three hundred yards. As before, be sure to let your eyes stop at specific points in the search rather than just sliding by them.

When this search is completed, maintain a watch over the area to detect any enemies. This watch should be a cross between the thirty-second search and the longer fifty-yard-at-a-time search outlined above.

Quick movements are most apt to give an enemy—or you—away. But you should be looking for reflections or improper camouflage technique as well. Listen for unusual sounds: Though the location of a sound is often hard to pinpoint, it can alert you to the presence of an enemy.

Since a trained enemy will not remain exposed for long, take note of his surroundings. You may wish to hold your fire until you can ascertain how many of the enemy there are. If so, it is essential you mentally mark the spot in which he is standing so that you can locate his position later. The best way to do this is to use either an aiming or reference point.

An aiming point is an object behind which the enemy is hiding or in which he is concealed that can be penetrated by your bullets. The aiming point is easier to find—as well as hit—than the reference point.

A reference point would be some object near an enemy's position (a tree, for example). With the reference point, it is necessary to estimate how far behind, in front of, or to the side of the object the enemy is.

Be sure your aiming point or reference point is easy to distinguish. You must remember *which* bush or boulder the enemy was near. It can be difficult and time-consuming to explain to a companion just what point you were looking at.

Though an enemy will rarely be engaged at ranges beyond three hundred yards, it is almost essential that you have as accurate an idea as possible as to how far away from you he is. Ideally, you would have a scoped rifle or range finder capable of giving the precise distances. In combat, this will rarely happen and the scope/range finder may not be operative.

A lot can be done to improve your judgment of ranges simply by stepping off one hundred yards from time to time to see just how far that distance really is. Purchasing a range finder and using it to check your guesses will also increase your skill.

When you estimate how far one hundred yards is, you should try to mentally "step off" groups of one-hundred-yard increments in the area in which you are looking. Again, having stepped them off or used a range finder beforehand to check yourself will be a real plus.

At ranges of five hundred yards or less, it is possible for most people to make very accurate estimates after a little practice. It is, of course, more difficult to be accurate at ranges over five hundred yards. One useful way of estimating such long ranges is to pick a point halfway to the target and estimate the distance in one-hundred-yard increments. Double the halfway figure, and you have your distance from the target.

Objects can sometimes appear to be closer than they are. Factors which can cause this distortion of distance perception include:

- You are looking across a depression which is hidden from view.
- You are looking down from high ground.
- You are looking down a straight, open path, road, or railroad track.
- You are looking over a uniform space such as a river, snow, desert sand, or a wheat field.
- The sun is at your back.
- The "target" is silhouetted by light, color, or outline.
- The air is very clear (as it is at high altitudes).
- The entire "target" is visible.

Objects can also appear farther than they actually are for the following reasons:

- You can only see part of the target.
- The target is small in relation to its surroundings.
- You are looking over a depression which is visible.
- You are looking to higher ground.
- Your field and length of vision are small (in a confined street, a draw, or a forest trail).
- You are viewing an area in poor light.
- The sun is in your eyes.
- The air is not clear due to dust, rain, or snow.
- The enemy blends well with the terrain.

A trick to use to gauge a person's distance from you is to judge the distance by the size something (of known size) appears to be. This is the normal, everyday way of judging distances on the highway or on the street.

This method of ranging can be developed and, with practice, become very accurate. Count off one hundred yards and see what size a person appears at that distance. Fix this apparent size in your mind so that you can estimate the range an enemy is from you at a later date.

Another way to estimate distance, which is especially suited to defense, is to make a sector sketch of the ranges of objects in your area. This can be done mentally or on paper. The sketch allows you to estimate the range of an enemy by his proximity to a landmark of known range. A landmark will also help you point out an enemy's position to others in your group, or it can be used as a reference point so that you can relocate the enemy if it is impossible to fire on him right away.

TIPS FOR IMPROVING ACCURACY OF FIRE

When firing on an enemy at some distance, there are a number of "tricks" you need to employ to improve the accuracy of your fire.

With stationary targets, you need to estimate the bullet drop involved with the distance the enemy is from you, and make some correction for windage.

Some types of scopes have been designed that will compensate for bullet drop. With these scopes, estimating bullet drop is not a problem if the scope makes the right adjustments for your ammunition, the scope is working, and you have enough time to get it into action. Though suited for sniper work, the compensating scope often is not fast enough to be used in combat situations.

Iron sights, "red dot" sights, or simple scopes can be quickly brought into play. It is critical, however, to have the gun zeroed in at the proper yardage. With the AR-15, the battle zero for the iron sights is generally 250 yards, and this should be made on the rear one- to three-hundred-yard aperture. (The "L" rear leaf is for three hundred- to five hundred-yard targets). You must give some thought as to what your maximum range might be. If you are in an area where the maximum range is three hundred yards, zeroing in at one hundred yards might make sense. Likewise, if you might be firing out to six hundred yards, a zero of four hundred yards with the "L" aperture could be the wisest move.

Adjusting a rifle to a two hundred fifty-yard range sounds like a chore. It is not as hard as it sounds, though, if you first zero in by removing the upper receiver/barrel and rough them in so that they point in the same direction. Zero in by firing the rifle at thirty yards first; make the fine adjustments by firing at two hundred fifty yards. (Thirty yards is the approximate point in the ballistic curve at which the bullet will be at two hundred fifty yards. If you cannot zero in at two hundred fifty yards, thirty yards would probably give you a fairly good zero for combat.)

BALLISTIC CHARTS AND RANGES

If you study the following ballistic charts, you will see that a battle zero of two hundred fifty yards means that the bullet will be impacting within five inches up or down with any shot you make at under three hundred yards. Since three hundred yards is considered the "combat maximum," you will probably need to give little thought to the bullet drop—except in cases where only a small portion of an enemy's body is exposed to fire. Bullet drop does become very important with shots over three hundred yards.

To make hits over three hundred yards, you obviously need to judge how much hold-over you need to take when shooting at the various ranges. To do this you must have a good idea of the range and the ballistic behavior of the round used.

It should also be noted that when you fire up or down a steep slope, the bullet only drops over the horizontal range it covers rather than the actual distance it travels.

I have often spoken to survivalists who plan to defend an area by using five-hundred-yard or longer shots (usually with a rifle chambered for .308). While this sounds good on paper, a little study of the effectiveness of shots beyond three hundred yards, coupled with the fact that most hits occur within one hundred yards, seem to indicate this defense is not feasible.

With iron sights, the chance of hitting a human target beyond three hundred yards is small unless you are firing a mounted machine gun.

In terms of safety, it is interesting to note that the 5.56mm has a maximum range of three thousand yards. If you fire the rifle into the air, you may accidentally hit someone from that distance, but the ballistic arc to get the bullet there makes it impossible to aim the rifle to obtain hits at these ranges.

The ability of a shooter with an AR-15 to

achieve reasonable hits beyond a maximum range of 715 yards is limited by this ballistic curve. (With a rifle chambered for 7.62 NATO, the curve limits realistic hits to 760 or fewer yards.) Someone planning on making hits beyond these ranges—even with a scoped rifle—is kidding himself unless he has a .50 sniper rifle or an M60 machine gun.

With a scope and steady rest, the chances of a hit are better, though the maximum limit of the range of 715 yards is still the same. The difference is that a scope can improve your ability to hit what you are shooting at. A scope that has a built-in bullet-drop compensator improves the odds somewhat for ranges out to six hundred yards. Automatic fire is not effective at great ranges unless you have an awful lot of it. With an AR-15 rifle, several semiauto shots at long range (rather than a burst of fire) are more likely to hit a target. That, however, is only true *if* you put in a lot of practice, have a good scope and accurate ammunition, and can accurately determine windage and range compensation. And beyond three hundred yards most shooters cannot compensate for all those "ifs."

But what if you *have to* fire at a target from that distance? Even though the chances of hitting it may be minimal, let us look at a few things that can be done to improve your accuracy.

Crosswinds between you and the target can do more to make you miss long-range targets than any other factor. Beyond one hundred yards, bullet displacement can determine whether or not you will hit the target or only alert someone to your presence.

WIND SPEED

Just as it is necessary to be able to judge the correct distance to compensate for bullet drop, so too, it is necessary to have an accurate idea of the wind speed to compensate for the deflection it will exert on a bullet.

Since most of us do not carry wind-speed equipment in our hip pocket, it is necessary to find the wind speed by estimating it. One way to do so is to note the angle of deflection the wind creates on some light object when it hangs or falls. When this angle is found, you can divide the number of degrees by four and come up with a close approximation of the wind speed.

If, for example, a flag or cloth is blowing in the wind, notice how high it is from the post. If the angle between the cloth and the pole below it is

sixty degrees, you divide sixty by four to discover that the wind speed is about fifteen miles per hour.

You could also drop a cloth or paper from near your shoulder. Note how far it blows out from a straight line between where you released it and the point where it landed. Estimate the angle created between your arm and body. Divide the number of the angle by four, and you will have the wind speed.

A quicker way of estimating wind speed is to use the following guides:

- A breeze under three miles per hour (mph) can hardly be felt, but it will cause smoke to drift.
- Three-to-five-mph breezes can be lightly felt on the face.
- Breezes of five to eight mph will keep tree leaves in constant motion.
- Eight to twelve mph will raise dust or blow loose papers.
- A twelve- to fifteen-mph wind will cause small trees to sway.

The following chart shows the displacement that will take place if the wind is blowing from your side. A wind blowing into your face or from your back will cause no noticeable displacement. (Getting the wind to your front or back is a good move if you are able to do so when sniping at an enemy.)

If a wind is coming toward your front or back at an oblique angle, give it a fractional value (roughly one third to two thirds of the full side wind value), depending on whether it is coming closer to being from the side or from your front or back.

EFFECTIVENESS AND RANGE IN COMBAT

How often will you hit the target if you can compensate for bullet drop and windage? The militaries of the world have tried to answer that problem through extensive tests. If you can use your rifle and compensate for windage and bullet drop, you will probably hit what you aim at 50 percent of the time at 360 yards or less, and one out of seven times out to a maximum range of 520 yards with the 5.56mm. Beyond that distance, the 1-in-12 twist AR-15 is too unpredictable for use. (The new faster twists and heavier bullets can score more hits at longer ranges.)

Those who feel that the .308/7.62mm NATO is better suited for long-range combat will be disappointed to find that it only improves things slightly: A good rifleman can score hits 50 percent of the time at ranges under 370 yards, or he can hit a target one out of seven tries out to 580 yards. When you compare the slightly increased range of the 7.62mm over the 5.56mm to the weight of the two weapons and their ammo, it is easy to see why many countries have switched to the 5.56mm. The superiority of the 5.56mm is especially evident in normal combat where ranges rarely extend past three hundred yards, and troops are either often poor marksmen or do not have time to aim carefully.

The amount of wind deflection is dependent on the ballistic coefficient of a bullet, not its size or weight. Many 7.62mm bullets have ballistic coefficients about twice those of their 5.56mm counterparts. But that is changing. New, heavier bullets are being developed for the 5.56mm that rival the performance of the 7.62mm. The new bullets will have as little deflection by the wind as the larger bullet now enjoys.

One of these new bullets is seen on the new NATO SS109 ammunition (designated the M855 by the U.S. military. The 62-grain, boat-tail bullet of the SS109 has a steel penetrator insert. This round can defeat helmets and body armor out to 880 yards and is thereby capable of providing lethal fire 44 percent farther than the 7.62mm NATO (according to extensive NATO tests).

Does a bullet retain enough energy beyond three hundred yards to be used in combat?

There are several factors to keep in mind when studying charts to come up with answers to such questions. One consideration is that a rifle bullet normally kills because of its speed, not because of size. When a rifle bullet drops below the 2,000 fps cutoff point, it becomes an inferior pistol bullet for all practical purposes.

The effectiveness of a rifle bullet which has dropped to pistol speeds depends on how many foot pounds it retains and how quickly it can transfer them to its flesh-and-blood target. One way to judge the effectiveness is to compare it to a bullet that has a reputation—good or bad—in combat.

With full-metal-jacket bullets, this comparison is more accurate with AR-15s that have a 1-in-12 twist than it is with the faster twists since the bullet is more apt to tumble. With the newer fast twists, solid point ammunition will not "dump" its energy efficiently; fast twists can be effective with hollow- or soft-point bullets (such options, though, are not available to military shooters).

Keep in mind that all bullet charts will give different results from those achieved in actual shooting. There are any number of variables that can create large changes in bullet effectiveness. Charts are useful guides for bullet comparisons and general information.

With these points in mind, we can see that the AR-15 loaded with a maximum load round will be effective as a high-velocity round out to three hundred yards. (If you reload, approach the maximum loads with caution.) Those who understand the principles of high-velocity wounds know that the bullet will be quite lethal within that range.

Beyond the point where the bullet sinks below 2,000 feet per second, it will lose much of its lethality and will behave like a small-caliber pistol bullet. At this point, the effectiveness of the bullet is determined by its energy and ability to quickly shed this energy in its target. Comparing the .22 bullet's energy to that of pistol rounds will give a very rough idea of what the bullet is capable of at ranges beyond three hundred yards.

BULLET DATA

55-GRAIN (FMJ) .223 WITH 250-YARD ZERO

	Muzzle	100 yds.	200 yds.	300 yds.	400 yds.	500 yds.	600 yds.
Deviation (Dev.) from 0	-1.5"	1.5"	2.5"	-3"	-17.8"	-45"	-90"
Velocity	3100 fps	2640 fps	2226 fps	1859 fps	1544 fps	1277 fps	1093 fps
Energy	1174 ft lbs	852 ft lbs	605 ft lbs	422 ft lbs	291 ft lbs	199 ft lbs	146 ft lbs

55-GRAIN (FMJ) .223 WITH 250-YARD ZERO (MAXIMUM LOAD)

	Muzzle	100 yds.	200 yds.	300 yds.	400 yds.	500 yds.	600 yds.
Dev. from 0	-1.5"	1"	2.8"	-4.1"	-18.3"	-41.6"	-87"
Velocity	3240 fps	2877 fps	2543 fps	2232 fps	1943 fps	1679 fps	1455 fps
Energy	1282 ft lbs	1011 ft lbs	790 ft lbs	608 ft lbs	461 ft lbs	344 ft lbs	259 ft lbs

55-GRAIN (FMJ) .223 WITH 300-YARD ZERO

	Muzzle	100 yds.	200 yds.	300 yds.	400 yds.	500 yds.	600 yds.
Dev. from 0	-1.5"	4.8"	6"	0"	-16.4"	-47"	-98"
Velocity	3100 fps	2640 fps	2226 fps	1859 fps	1544 fps	1277 fps	1093 fps
Energy	1174 ft lbs	852 ft lbs	605 ft lbs	422 ft lbs	291 ft lbs	199 ft lbs	146 ft lbs

55-GRAIN (FMJ) .223 WITH 300-YARD ZERO (MAXIMUM LOAD)

	Muzzle	100 yds.	200 yds.	300 yds.	400 yds.	500 yds.	600 yds.
Dev. from 0	-1.5"	4.4"	5.6"	0"	-15.1"	-43.8"	-91.7"
Velocity	3200 fps	2732 fps	2307 fps	1931 fps	1604 fps	1324 fps	1126 fps
Energy	1251 ft lbs	912 ft lbs	650 ft lbs	456 ft lbs	314 ft lbs	214 ft lbs	155 ft lbs

60-GRAIN .223 WITH 250-YARD ZERO

	Muzzle	100 yds.	200 yds.	300 yds.	400 yds.	500 yds.	600 yds.
Dev. from 0	-1.5"	3.25"	2.9"	-4.3"	-20"	-48"	-92.5"
Velocity	3000 fps	2633 fps	2296 fps	1991 fps	1716 fps	1467 fps	1260 fps
Energy	1199 ft lbs	924 ft lbs	703 ft lbs	528 ft lbs	392 ft lbs	287 ft lbs	211 ft lbs

60-GRAIN .223 WITH 250-YARD ZERO (MAXIMUM LOAD)

	Muzzle	100 yds.	200 yds.	300 yds.	400 yds.	500 yds.	600 yds.
Dev. from 0	-1.5"	1.9"	2.9"	-2"	-14.8"	-37.6"	-74"
Velocity	3200 fps	2820 fps	2467 fps	2145 fps	1854 fps	1596 fps	1365 fps
Energy	1365 ft lbs	1060 ft lbs	811 ft lbs	613 ft lbs	458 ft lbs	340 ft lbs	248 ft lbs

69-GRAIN .223 WITH 250-YARD ZERO (MAXIMUM LOAD)

	Muzzle	100 yds.	200 yds.	300 yds.	400 yds.	500 yds.	600 yds.
Dev. from 0	-1.5"	2"	3.5"	-4"	-12.4"	-35"	-71"
Velocity	3000 fps	2724 fps	2462 fps	2214 fps	2004 fps	1786 fps	1555 fps
Energy	1379 ft lbs	1137 ft lbs	929 ft lbs	751 ft lbs	615 ft lbs	488 ft lbs	370 ft lbs

OTHER CALIBERS (FOR COMPARISONS)

150-GRAIN .308 WITH 250-YARD ZERO

	Muzzle	100 yds.	200 yds.	300 yds.	400 yds.	500 yds.	600 yds.
Dev. from 0	-1.5"	3.5"	3"	-4.5"	-20"	-47"	-90"
Velocity	2820 fps	2593 fps	2396 fps	2210 fps	2035 fps	1869 fps	1714 fps
Energy	2648 ft lbs	2240 ft lbs	1913 ft lbs	1628 ft lbs	1379 ft lbs	1164 ft lbs	979 ft lbs

180-GRAIN .308 WITH 250-YARD ZERO

	Muzzle	100 yds.	200 yds.	300 yds.	400 yds.	500 yds.	600 yds.
Dev. from 0	-1.5"	4"	3.25"	-4.8"	-21.75"	-49.7"	-90"
Velocity	2600 fps	2393 fps	2198 fps	2015 fps	1842 fps	1682 fps	1535 fps
Energy	2703 ft lbs	2290 ft lbs	1932 ft lbs	1623 ft lbs	1357 ft lbs	1131 ft lbs	942 ft lbs

230-GRAIN .45 ACP

	Muzzle	50 yds.	100 yds.
Velocity	1000 fps	938 fps	888 fps
Energy	335 ft lbs	308 ft lbs	284 ft lbs

.22 LONG RIFLE (40-GRAIN)

	Muzzle	100 yds.
Velocity	1255 fps	1016 fps
Energy	140 ft lbs	92 ft lbs

WIND DEFLECTION ON 55-GRAIN .223 BULLET (3240 FPS MUZZLE VELOCITY)

	Muzzle	100 yds.	200 yds.	300 yds.	400 yds.	500 yds.	600 yds.
4 MPH WIND	0	.44"	1.6"	4.0"	8.0"	13.6"	20.0"
8 MPH WIND	0	.88"	3.5"	7.9"	16.0"	27.2"	40.0"
10 MPH WIND	0	1.0"	4.4"	10.6"	20.1"	34.0"	53.3"
20 MPH WIND	0	2.2"	8.9"	20.0"	40.2"	68.0"	100.0"

WIND DEFLECTION ON 69-GRAIN .223 BULLET (3000 FPS MUZZLE VELOCITY)

	Muzzle	100 yds.	200 yds.	300 yds.	400 yds.	500 yds.	600 yds.
4 MPH WIND	0	.4"	1.5"	2.6"	5.7"	10.7"	15.5"
8 MPH WIND	0	.8"	2.9"	5.2"	11.3"	21.4"	31"
10 MPH WIND	0	.9"	3.7"	7.0"	14.2"	27.9"	41"
20 MPH WIND	0	2.0"	7.4"	14.0"	28.1"	56.0"	82"

WIND DEFLECTION ON 180-GRAIN .308 BULLET (3240 FPS MUZZLE VELOCITY)

	Muzzle	100 yds.	200 yds.	300 yds.	400 yds.	500 yds.	600 yds.	700 yds.
4 MPH WIND	0	.26"	1.3"	2.9"	4.2"	8"	12"	17"
8 MPH WIND	0	.53"	2.8"	5.7"	10.0"	16"	25"	33"
10 MPH WIND	0	.6"	3.5"	7.2"	13.0"	20"	29"	41"
20 MPH WIND	0	1.3"	7.0"	14.0"	26.0"	40"	58"	82"

(Note: Actual figures for your gun and ammunition will vary greatly from those shown above. This table should be used for reference only.)

One good pistol round to compare it to is the .45 ACP. Even with a full metal jacket and a rounded point (which tend to minimize the round's abilities to damage living tissue), most authorities consider the .45 ACP to have a "knockdown" value of 90 percent. In other words, a shot to the torso will take an enemy out of the fray, though not necessarily kill him, about 90 percent of the time. The .45 ACP creates this effect with 335 foot pounds of energy at its muzzle and a wide diameter.

Another good round for comparison is the small .22 Long Rifle. The .22 LR is deadly, but it is not a noted fight-stopper with its miniscule 80 to 100 foot pounds of energy. At its muzzle, the .22 LR comes close to the 108 foot pounds of energy which most military authorities consider the absolute minimum amount of energy needed to wound a man enough to take him out of battle.

Referring to the charts, notice that the 55-grain 5.56mm loaded to its maximum will be equivalent to the .45 ACP as a fight-stopper out to three hundred yards provided it dumps its energy by tumbling or expanding. (With a full metal jacket and fast rifling, the bullet will not have as much knock-down ability.) The 60-grain bullet would be equivalent to the .45 ACP another one hundred yards beyond the 55-grain bullet, and the new 69-grain commercial bullets (currently being marketed by the Sierra Bullet Company under the trade name of "Matchking") could extend that range out to six hundred yards.

A quick look at the gyroscopic stability chart shows that a 60-grain or larger bullet is extremely unstable when fired from a 1-in-12 twist. This makes it inaccurate, but deadly, with slow-twist barrels. Such a round might be used in close-quarter combat by those limited to full-metal-jacketed bullets by choice or regulation if they have access to an older AR-15 model.

Using the military 108 foot pounds as the minimum needed to remove a soldier from combat and remembering that fast twists require the use of an expanding bullet, the 5.56mm can be of use out to six hundred yards. The new 69-grain commercial bullets can theoretically engage targets beyond that range. The stable, armor-piercing bullet SS109 would have the energy but would not dump its energy quickly enough to be used at this range except with a machine gun, which might score multiple hits, or in the dubious task of engaging vehicles with the AR-15.

This six-hundred-yard range is only useful on paper for the most part, however, since the chances of a hit are only 50 percent at 360 yards, and only one hit out of seven shots is probable at 520 yards. You certainly would never want to shoot from this range for anything other than harassment purposes or sniper work. Also consider that most people are not in areas where a six-hundred-yard target often presents itself. Normally such long ranges without branches or other obstructions between you and a potential target are rare, especially if you are to remain hidden from the enemy's return fire.

Your best bet is to stay within the three-hundred-yard limit, with even closer ranges being more of a sure bet if you are in a good defensive position. If you do not have mortars or heavy machine guns, long-range shots at an enemy who does would be suicidal.

Be aware of your abilities, and remember that most combat takes place within a distance of one hundred yards.

So far, we have assumed your target will be stationary, though rapid movement is in fact part of modern combat. This is one reason your AR-15 should have a 250-yard zero—you won't have time to make a lot of windage and drop adjustments. Instead, you will encounter a fleeting target and have to compensate with your aim.

When an enemy is moving, you must anticipate where he will be when the bullet gets to his location, rather than where he is when you fire. When he is moving laterally (parallel) to you, you should have a point of aim that is at the front edge of his body if he is within 250 yards of you. If he is beyond that distance, aim one body width in front of him. Double the lead space if he runs parallel to you. When an enemy moves toward you from an oblique angle (diagonally), use a lead that is one half of what you allowed for the lateral lead. When he runs from one location to another, aim at him when he slows down, which he will do at the beginning and end of each of his rushes.

At the other extreme of combat is close-quarters fighting, which often occurs in heavily wooded areas or dense urban environments. Under such circumstances, you may not have enough time to take careful aim and may have just enough time to fire before your enemy can fire at you! In such a case, it is essential that you be both quick and accurate.

The U.S. military developed several methods of

helping soldiers learn to efficiently hit targets at close range. One such method, known as quick fire or quick kill, requires the shooter to assume the same stance he would for normal shooting from the shoulder, looking over the rifle sights so that the barrel is parallel to where the shooter is looking.

One way to learn how to maneuver is to place tape over the rear and front sights and practice with a .22 adapter kit. Once targets can be hit at fifteen yards, you can start moving the target back.

Many shooters also find that the Armson O.E.G., Aimpoint, or Singlepoint scopes also help develop this skill so that targets can be hit quickly even after the scope is removed.

Full training techniques are provided in the U.S. Army's training text 23-71-1, "Principles of Quick Kill," available from Paladin Press (Box 1307, Boulder, CO 80306).